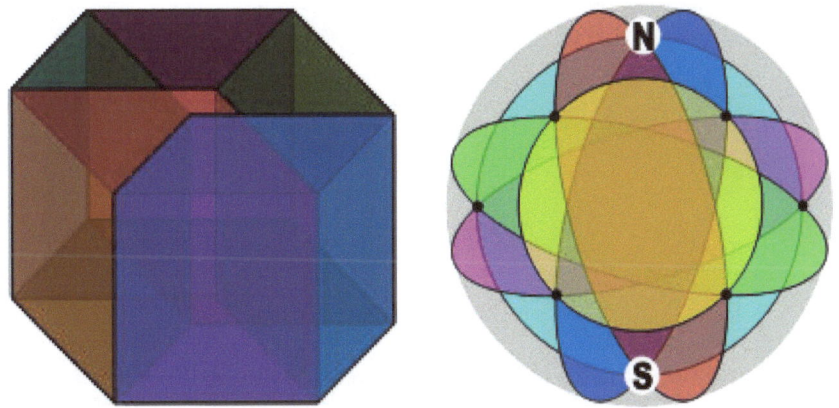

Full COLOR Illustrations of the
Fourth Dimension

Volume 1: Tesseracts and Glomes

Chris McMullen, Ph.D.

Chris McMullen, Ph.D.

Full COLOR Illustrations of the Fourth Dimension, Volume 1: Tesseracts and Glomes

Copyright © 2009 Chris McMullen, Ph.D.

All rights reserved. This includes the right to reproduce any portion of this book in any form.

Custom Books

Nonfiction / Science / Mathematics / Geometry

Nonfiction / Arts & Photography / Graphic Arts

ISBN: 1442141638

EAN-13: 9781442141636

Contents

Part I: Tesseracts (Four-Dimensional Hypercubes)

Mathematical definition of a tesseract	4
A tesseract and its eight bounding cubes	5
The points, edges, and square faces of a tesseract	6
Pairs of mutually orthogonal square faces	7
Pairs of mutually orthogonal cubes	8
The faces of the cubes bounding a tesseract	9
Viewing a tesseract from different angles	9
How to draw a tesseract in perspective	10
Dissection of a tesseract drawn in perspective	11
Tesseracts viewed from various perspectives	12
A cube reflects by rotating into the fourth dimension	13
Cubes rotating through the fourth dimension	14
Rotating a tesseract in four-dimensional space	15
Tesseracts rotating in the fourth dimension	16
A tesseract unfolding into the third dimension	17
The intersection of a hyperplane and tesseract	18
Dissection of a cross section of a tesseract	19
Three-dimensional cross sections of tesseracts	20
Rectangular hyperboxes in four dimensions	21

Part II: Glomes (Hyperspheres in the Fourth Dimension)

Mathematical definition of a glome	22
Six mutually orthogonal great circles on a glome	23
Four mutually orthogonal great spheres on a glome	24
Viewing a glome from different angles	25
Poles and compass directions in the fourth dimension	26
Hyperspherical coordinates in four dimensions	27
How to draw longitudes, latitudes, and hyperlatitudes	28
Circular longitudes, latitudes, and hyperlatitudes	29
Surfaces of longitude, latitude, and hyperlongitude	30
Three-dimensional lattice structure of a glome's hypersurface	31
A sphere rotating through the fourth dimension	32
Spheres rotating through the fourth dimension	33
Rotating a glome in four-dimensional space	34
Glomes rotating in the fourth dimension	35
The intersection of a hyperplane and glome	36
Slicing a glome into parallel adjacent spheres	37
The intersection of mutually orthogonal spheres	38
A hyperellipsoid in four-dimensional space	39

Part I: Tesseracts

(Four-Dimensional Hypercubes)

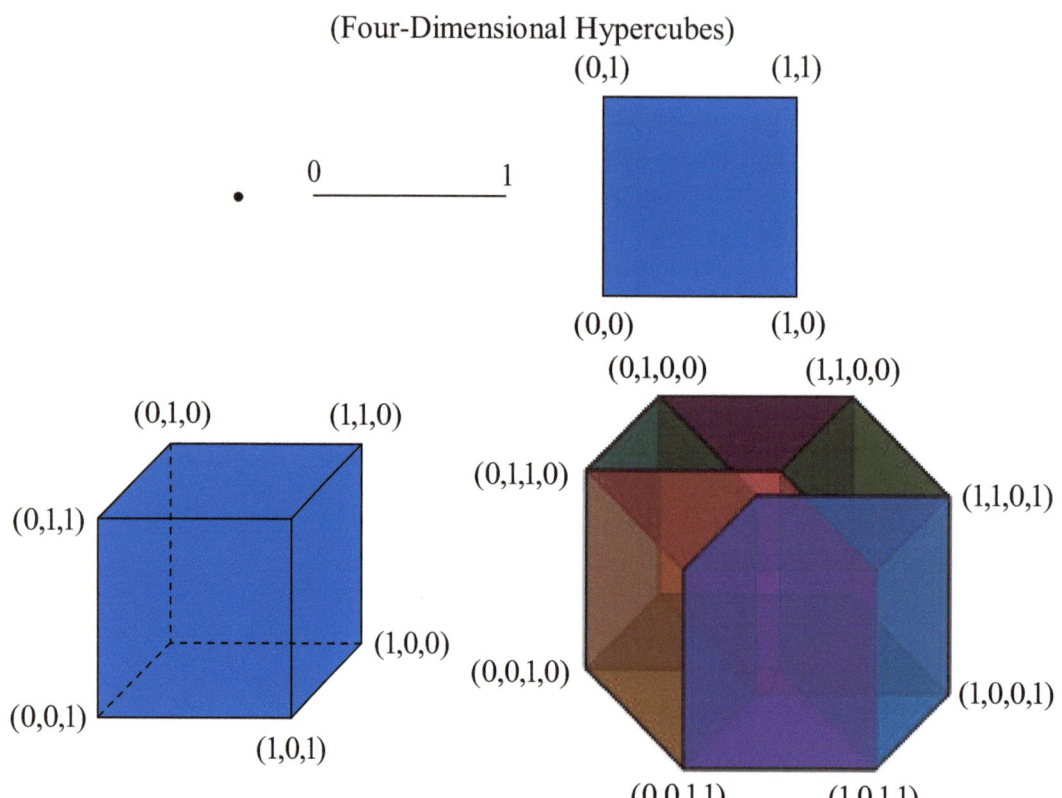

A **tesseract** is a 4D hypercube. The hypercube is a straightforward generalization of the cube to N-dimensional space as follows: In Cartesian coordinates – i.e. $(x, y, z, w,...)$ – the corners of an N-dimensional **unit hypercube** have binary components – i.e. each coordinate is either 0 or 1. These corners are connected by edges that run parallel to the Cartesian axes. A general hypercube can thus be constructed by rescaling (multiplying each coordinate by the same scaling factor), translating (moving the hypercube along a straight line), and rotating the unit hypercube. Some lower-dimensional unit hypercubes are:
- The 0D unit hypercube is a **point** at the origin.
- The 1D unit hypercube is a straight **line** with endpoints 0 and 1.
- The 2D unit hypercube is a **square** with endpoints (0,0), (0,1), (1,0) and (1,1).
- The 3D unit hypercube is a **cube** with endpoints (0,0,0), (0,0,1), (0,1,0), (0,1,1), (1,0,0), (1,0,1), (1,1,0), and (1,1,1).
- The 4D unit hypercube is a **tesseract** with endpoints (0,0,0,0), (0,0,0,1), (0,0,1,0), (0,0,1,1), (0,1,0,0), (0,1,0,1), (0,1,1,0), (0,1,1,1), (1,0,0,0), (1,0,0,1), (1,0,1,0), (1,0,1,1), (1,1,0,0), (1,1,0,1), (1,1,1,0), and (1,1,1,1).

Full COLOR Illustrations of the Fourth Dimension, Volume 1: Tesseracts and Glomes

The **tesseract** (a 4D hypercube) is a 4D generalization of the cube. All edges meet at right angles. 8 cubes bound the 4D volume inside. These bounding cubes are mutually **orthogonal** (perpendicular), just as squares can be perpendicular in 3D space. The bounding cubes are shaded with transparency. The bottom diagrams show the 4 front and 4 rear bounding cubes.

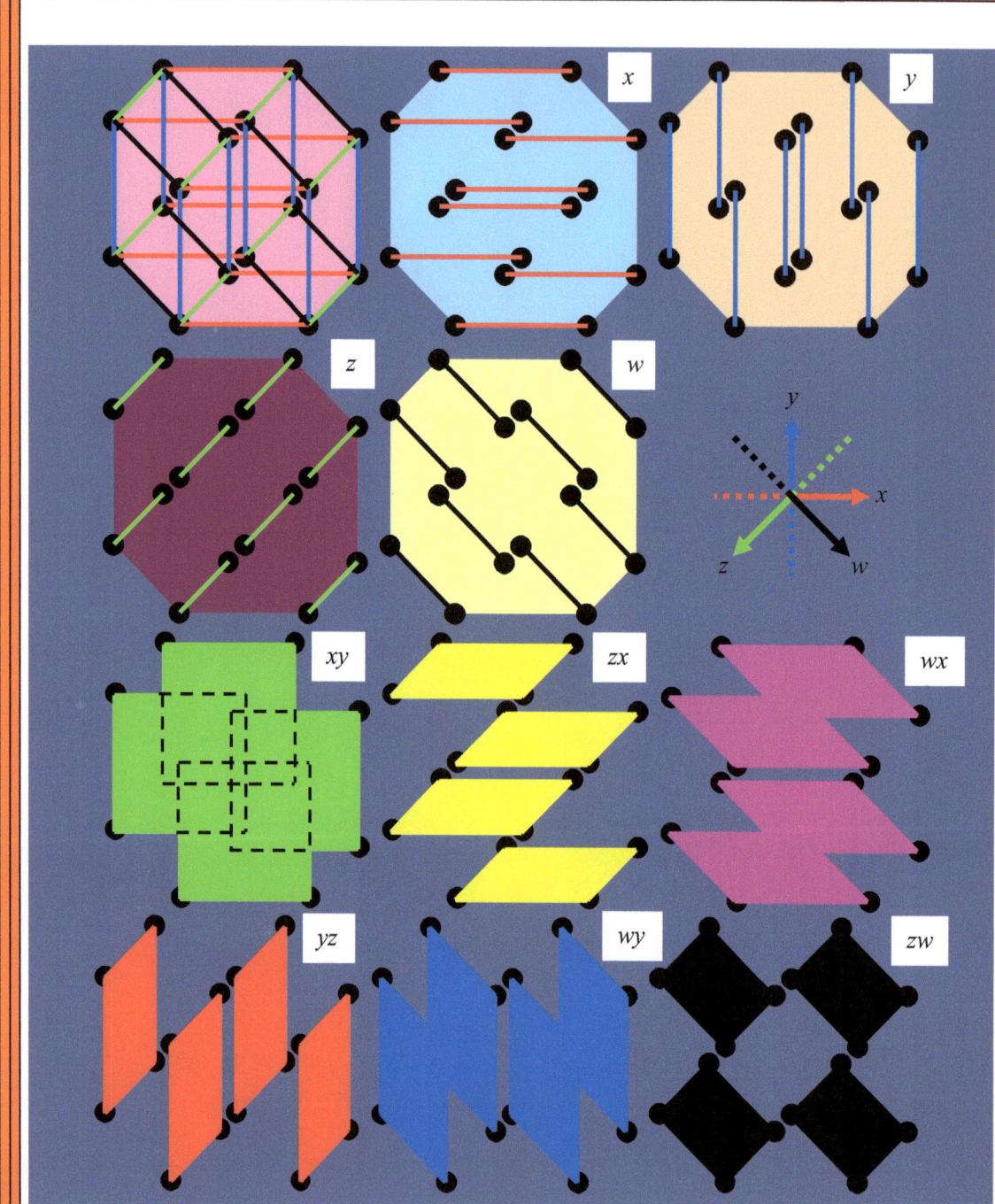

All together, there are **16 corners** (2 corners for each dimension multiplied together $= 2\times 2\times 2\times 2 = 16$), **32 edges** (a set of 8 parallel edges to join 8 pairs of corners along each of the 4 axes), **24 square faces** (a set of 4 parallel squares for each group of 4 corners lying in one of 6 possible planes), and **8 cubes** (not shown in this diagram) in one tesseract.

Full COLOR Illustrations of the Fourth Dimension, Volume 1: Tesseracts and Glomes

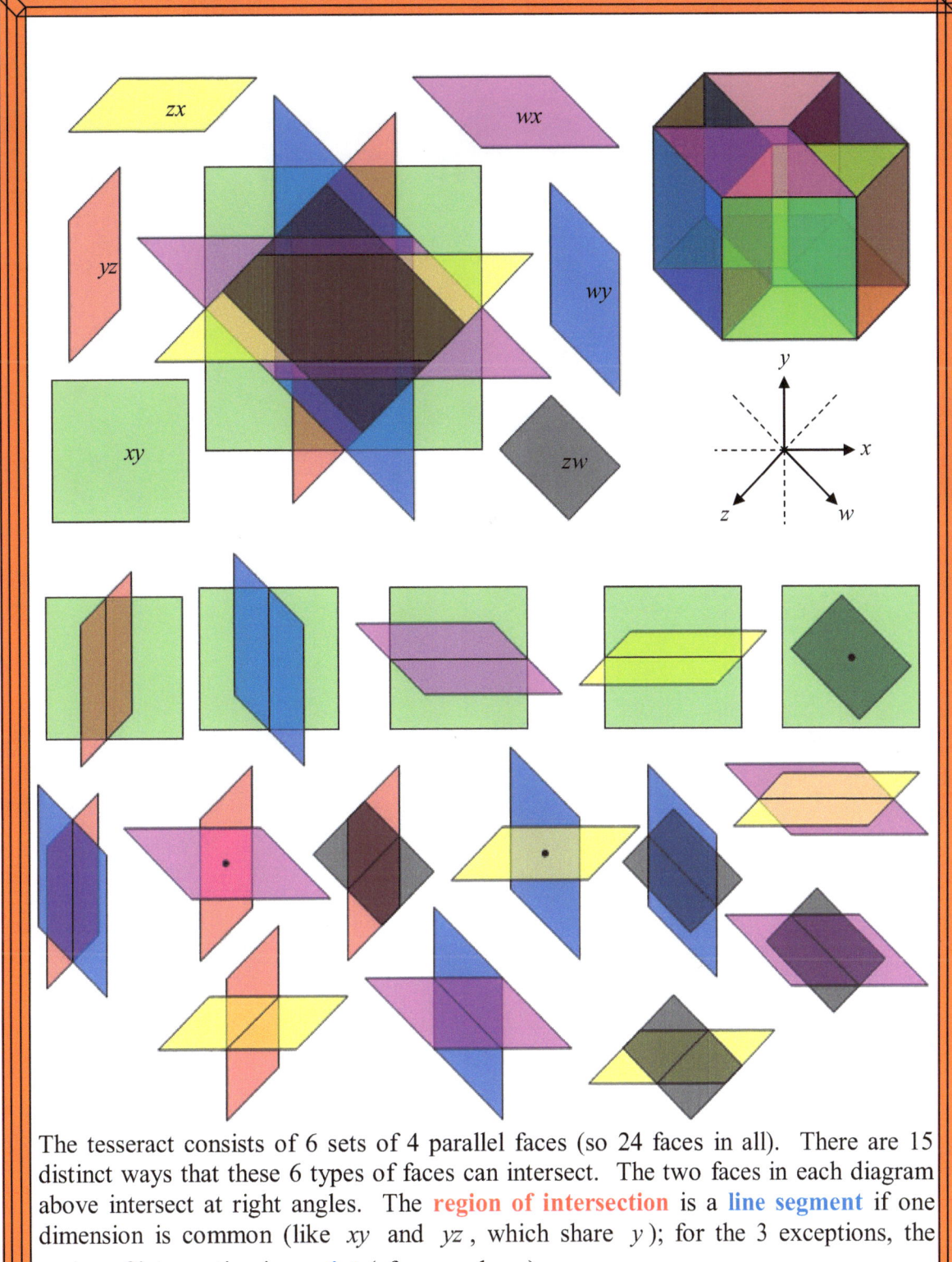

The tesseract consists of 6 sets of 4 parallel faces (so 24 faces in all). There are 15 distinct ways that these 6 types of faces can intersect. The two faces in each diagram above intersect at right angles. The **region of intersection** is a **line segment** if one dimension is common (like *xy* and *yz*, which share *y*); for the 3 exceptions, the region of intersection is a **point** (cf. *xy* and *zw*).

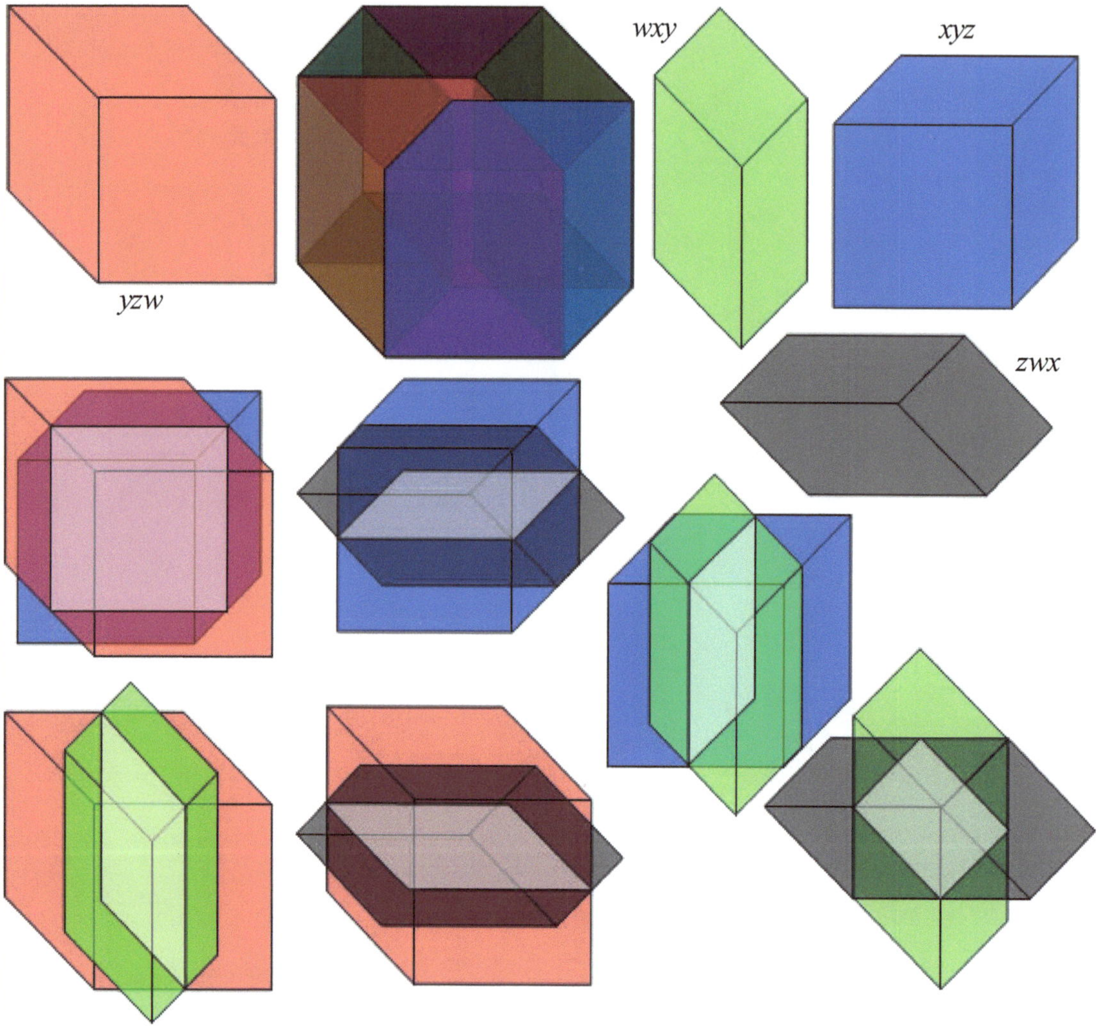

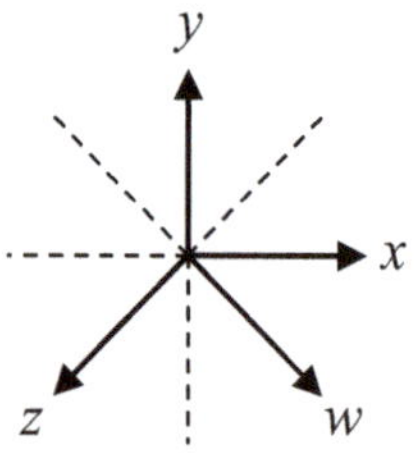

The tesseract consists of 4 sets of 2 parallel cubes (8 cubes all together). Each of the 4 kinds of cubes can intersect 6 distinct ways; the **region of intersection** is a **plane** parallel to one of the 6 types of square faces. Any two of these 4 different kinds of cubes are **mutually orthogonal**.

Full COLOR Illustrations of the Fourth Dimension, Volume 1: Tesseracts and Glomes

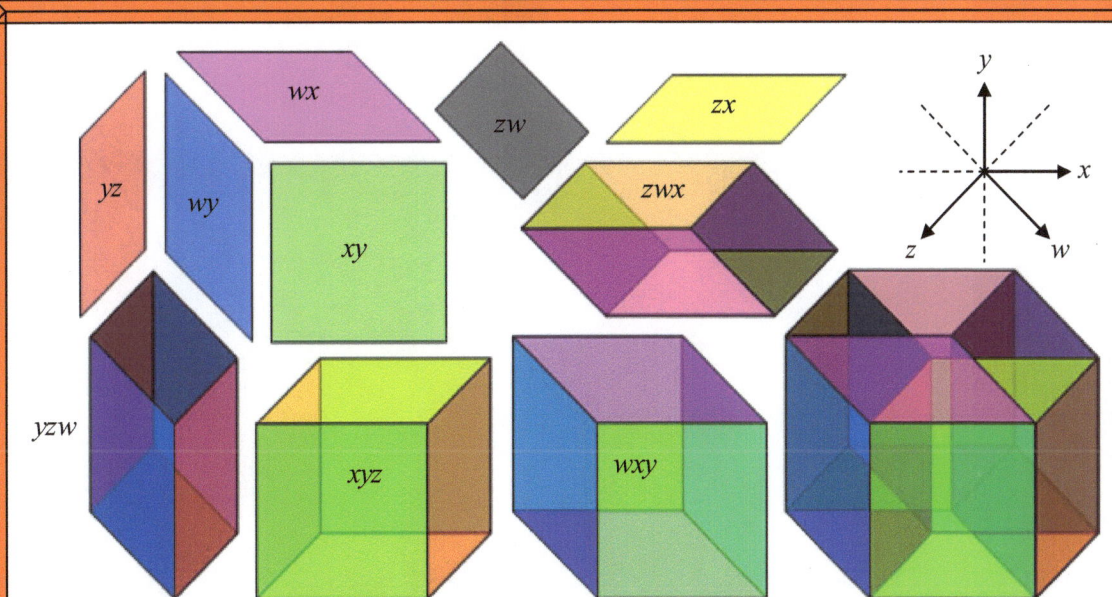

The tesseract has **8 bounding cubes**, where each cube has 6 square faces. Each face is shared by two cubes in the tesseract, so it consists of 24 (rather than 48) square faces. The relationships between the 6 distinct kinds of faces and 4 types of cubes are illustrated above.

Different **viewpoints** of the tesseract vary by amount of apparent depth and hyperdepth. The top right diagram shows a view along the main body diagonal. In this view, the tesseract appears as a regular octagon with a square drawn from each edge; a smaller octagon appears in the center. Compare to a cube viewed along the body diagonal, which appears as a hexagon with squares drawn from each edge. The bottom right view is analogous to viewing a cube where two sides, rather than three, are directly in view.

Chris McMullen, Ph.D.

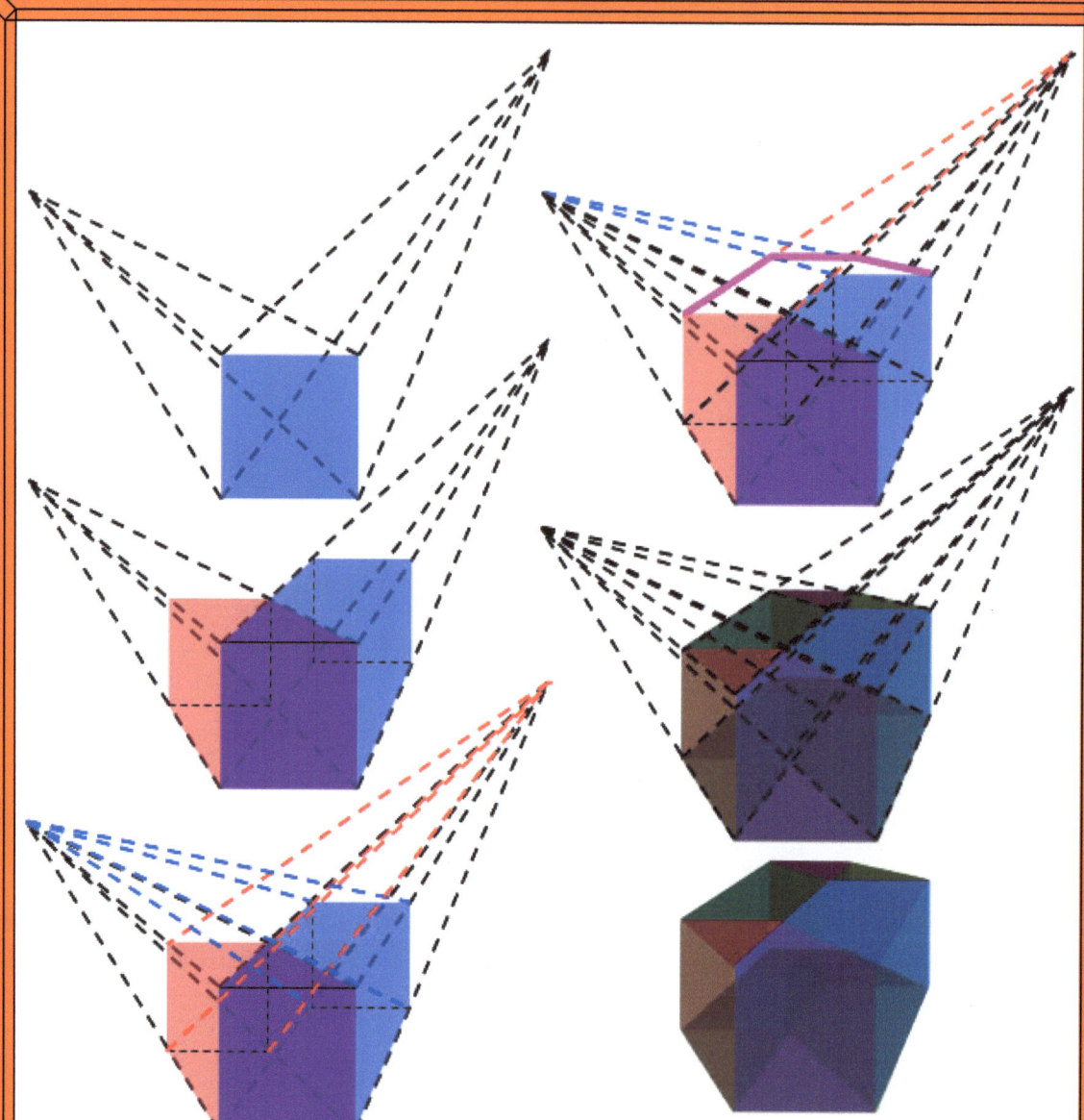

A tesseract can be drawn in **perspective** most easily if one set of faces is parallel to the plane of the paper; two perspective points are needed for this. **First**, choose the two perspective points; in principle, they may be far away (even further than the edge of the paper), in which case the effect is not as dramatic; they may also lie within the projected image of the tesseract, which produces more pronounced effects. **Second**, draw the front square face of the tesseract, which is parallel to the plane of the paper. **Third**, connect the corners of this face to the perspective points. **Next**, choose a depth and hyperdepth and use the perspective lines to extend the front square face into two orthogonal cubes. Draw four new perspective lines to the rear faces of these cubes. Use the new perspective lines to complete the outline of the tesseract. **Finally**, complete the interior of the tesseract by adding edges for the remaining (six) cubes.

Full COLOR Illustrations of the Fourth Dimension, Volume 1: Tesseracts and Glomes

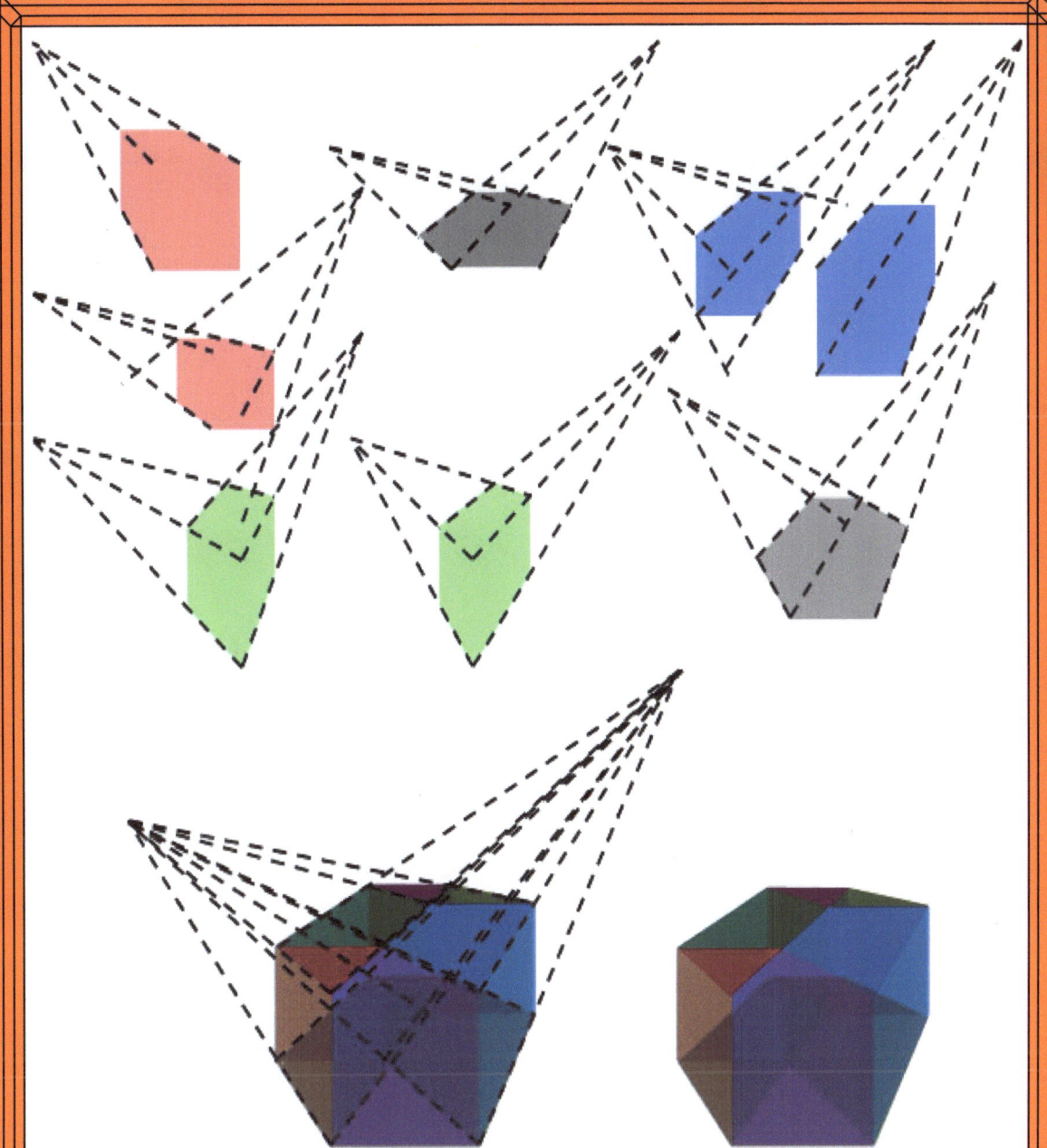

The 8 bounding cubes are shown for a tesseract drawn in **perspective**. The perspective of each cube is defined by three lines emanating from one of the two perspective points if it has both horizontal and vertical edges (where this perspective relates either to the depth or hyperdepth of the cube); otherwise, its perspective is determined by three lines emanating from both perspective points (since in this case the cube has both depth and hyperdepth). The tesseract has **width**, **height**, **depth**, and **hyperdepth**: Thus, if horizontal and vertical lines are employed to illustrate width and height, two perspective points are needed to achieve the four-dimensional perception of depth and hyperdepth.

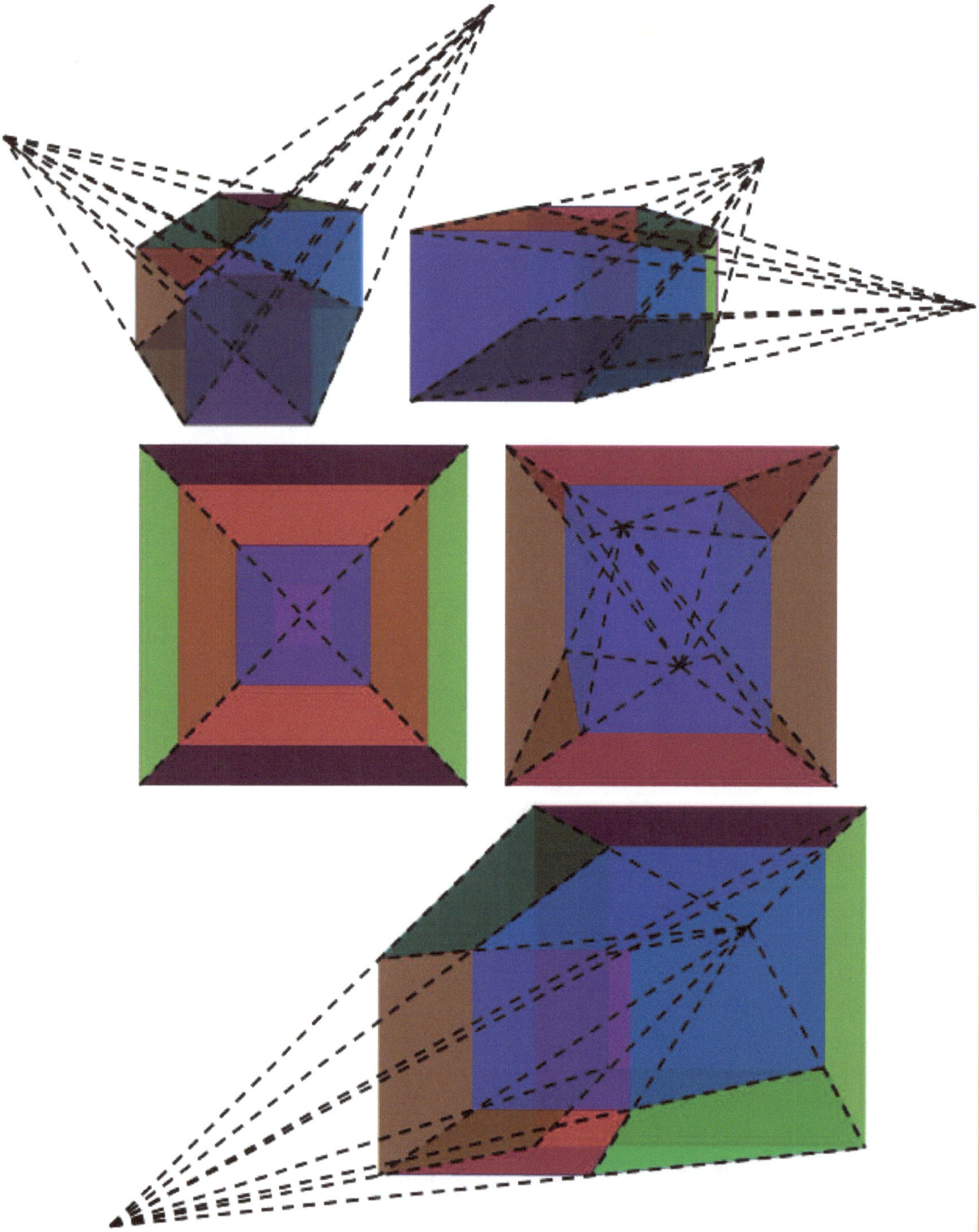

These tesseracts are drawn from **two perspective points**. For the tesseract on the left of the middle row, both perspective points appear in the center of the tesseract's two-dimensional projection.

Full COLOR Illustrations of the Fourth Dimension, Volume 1: Tesseracts and Glomes

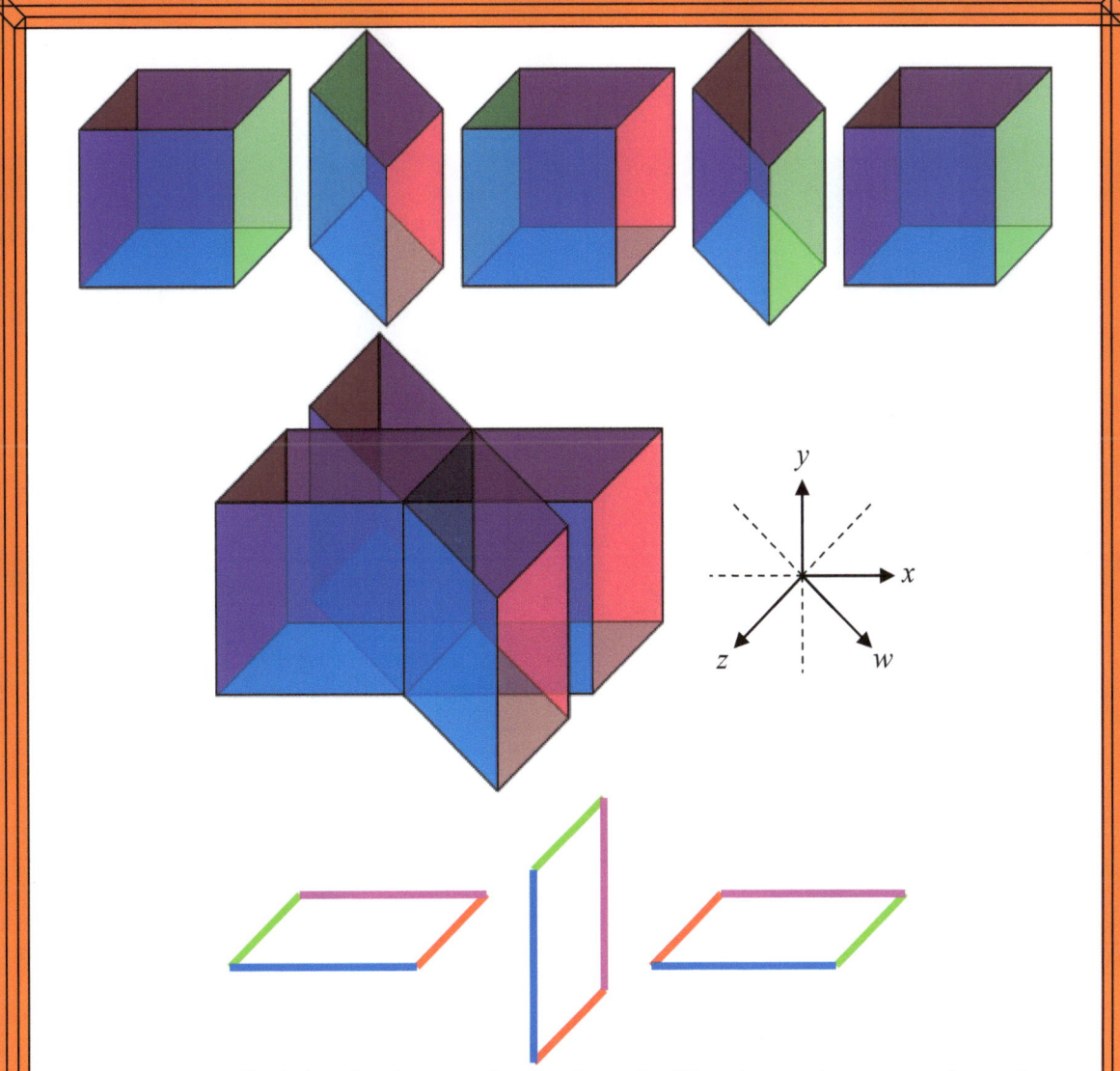

This cube, initially lying in the *xyz* hyperplane (a 3D subspace), rotates about the *yz* plane. Each rotation shown corresponds to a 90° **rotation in 4D space**. Each 90° rotation transforms the cube between the *xyz* and *yzw* hyperplanes; these two hyperplanes are mutually orthogonal and intersect at the *yz* plane (there is no other overlap). The 4 edges along *y* remain parallel to the *y*-axis throughout this rotation, and similarly for the 4 edges along *z*, while the other 4 edges rotate in the *wx* plane. A rotation about **180°** into the 4th dimension (i.e. the dimension that the cube does not have) **reflects the cube**. This is illustrated above as the red and green sides effectively swap over the course of a 180° rotation. The middle diagrams shows the four rotated images combined together, where again no two cubes overlap except at the common intersection at the *yz* plane (the green face in the center). The bottom image shows the analogous rotation of a square in the *zx* plane rotating about the *z*-axis into the third dimension, which results in a reflected 2D image after rotating 180°.

Each of the cubes illustrated above is **rotating** 360º in 90º intervals **through the fourth dimension** (i.e. the direction perpendicular to each of its dimensions). The four images of each diagram intersect only at the common plane of rotation, and otherwise do not overlap. Each cube effectively **reflects** about the plane of rotation through every **180º**.

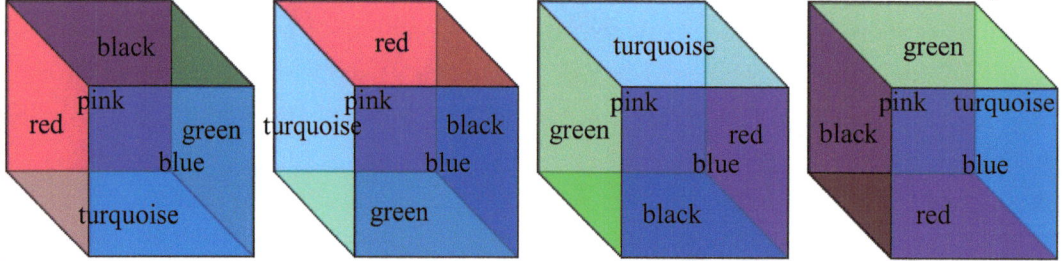

This cube is **not rotating into its fourth dimension**. This shows that not every rotation in 4D space is fully four-dimensional, and that a cube can rotate without reflecting.

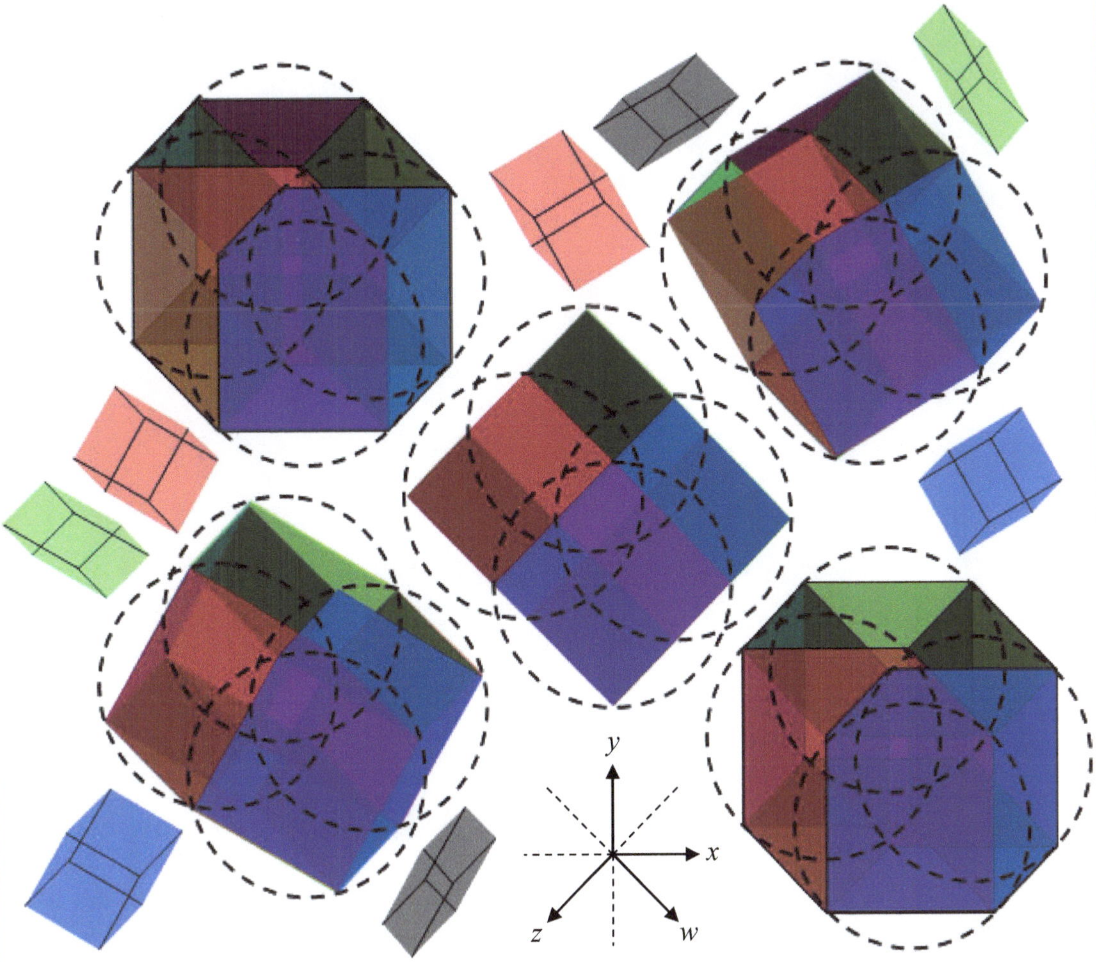

Full COLOR Illustrations of the Fourth Dimension, Volume 1: Tesseracts and Glomes

As this **tesseract rotates** about the *wz* plane, every point on the tesseract rotates in a circle in the *xy* plane. The 16 edges originally parallel to the *z*- and *w*-axes remain parallel to these axes, while the 16 edges in the *xy* plane rotate in the *xy* plane. The second, third, fourth, and fifth tesseracts (reading left to right and top to bottom) show rotations for 30°, 45°, 60°, and 90° relative to the first tesseract. Through 90°, the *xyz* and *wxy* cubes have returned to their original outlines, but 4 of their faces have rotated one position over, while the *yzw* cubes now occupy positions where the *zwx* cubes were and vice-versa. Two of the diagrams also show how the 4 distinct types of bounding cubes rotate as part of the transformation.

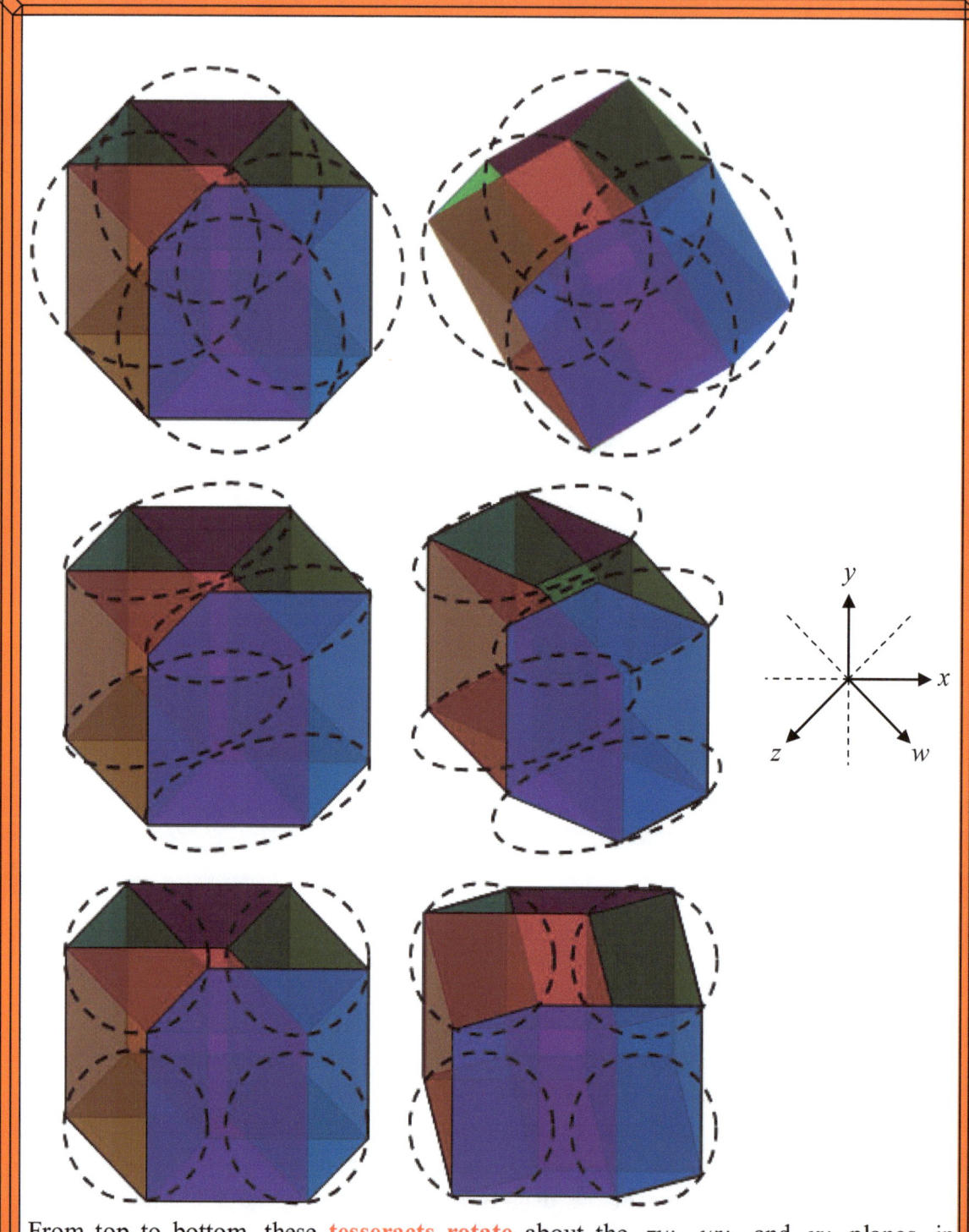

From top to bottom, these **tesseracts rotate** about the zw, wy, and xy planes, in which the zw, wy, and xy faces, respectively, maintain their orientation as the edges perpendicular to these faces rotate.

Full COLOR Illustrations of the Fourth Dimension, Volume 1: Tesseracts and Glomes

The **tesseract** can be **unfolded** one cube at a time just as a cube can be unfolded one square face at a time. Each cube in the figures above rotates through a 90º angle (into the dimension that is orthogonal to the cube). There are multiple ways to unfold the cube; the double cross that results here is one of a variety of possible final unfolded states.

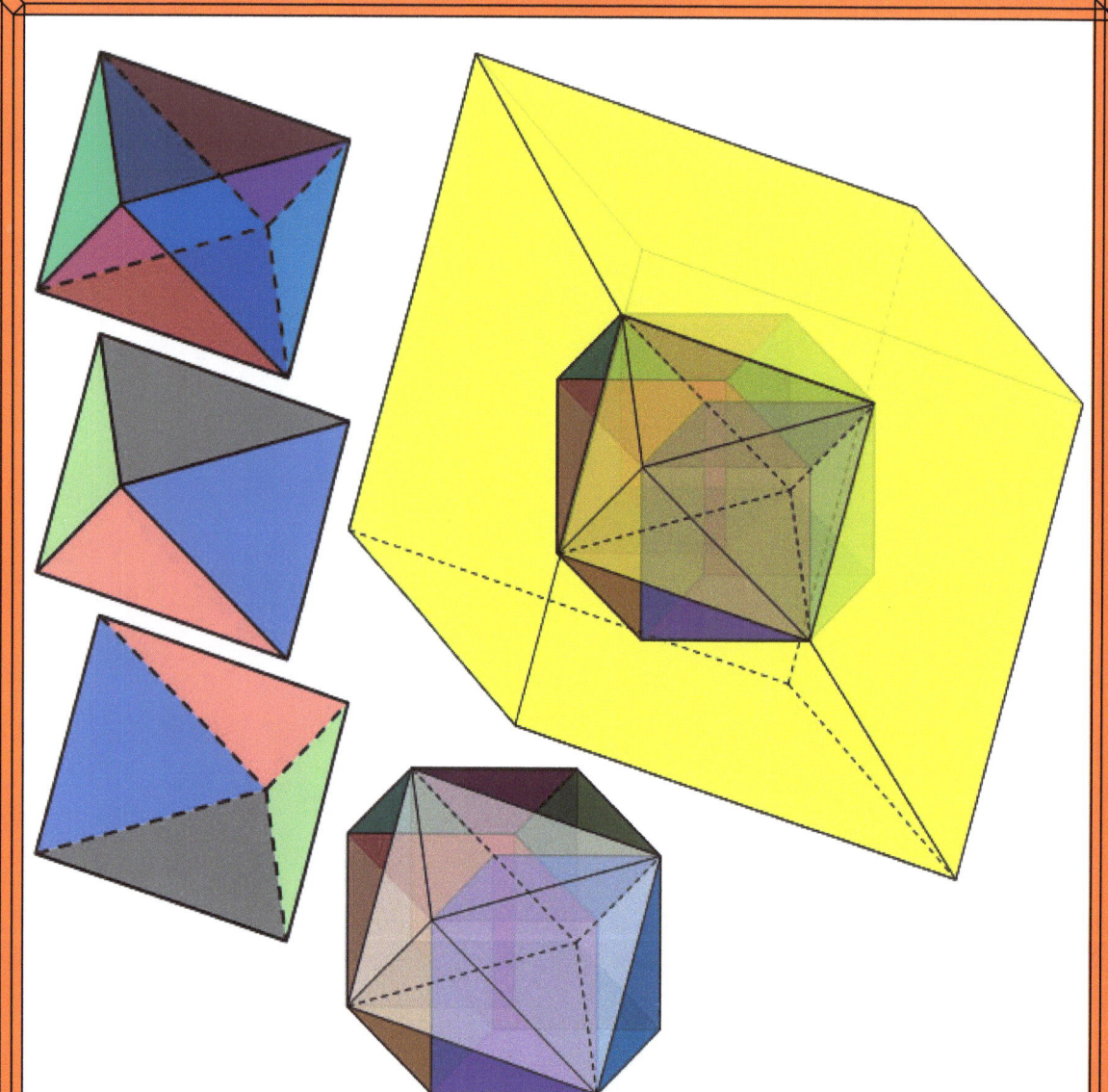

The **intersection** of a **3D hyperplane** (an infinite cube) and a **tesseract** can result in a **point**, **line segment**, or a **square face** as special cases, but more generally results in a **3D cross section**. The 3D cross section can have as few as **4 sides** (a **tetrahedron**) or as many as **8** (an **octahedron**). In the illustration above, the region of intersection is an octahedron. The hyperplane's faces are not parallel to any of the tesseract's squares, but instead the hyperplane slices through the tesseract along its long diagonal (analogous to slicing a cubic block of cheese with one stroke of a knife at an angle that includes two opposite corners, at such an angle that a hexagonal cross section results). Each of the 8 triangular faces of the octahedral cross section lies in one of the 8 bounding cubes of the tesseract. The images on the left show the front and back views of the octahedron as well as the composite object. Of course, the hyperplane is infinitesimally thin compared to the tesseract.

Full COLOR Illustrations of the Fourth Dimension, Volume 1: Tesseracts and Glomes

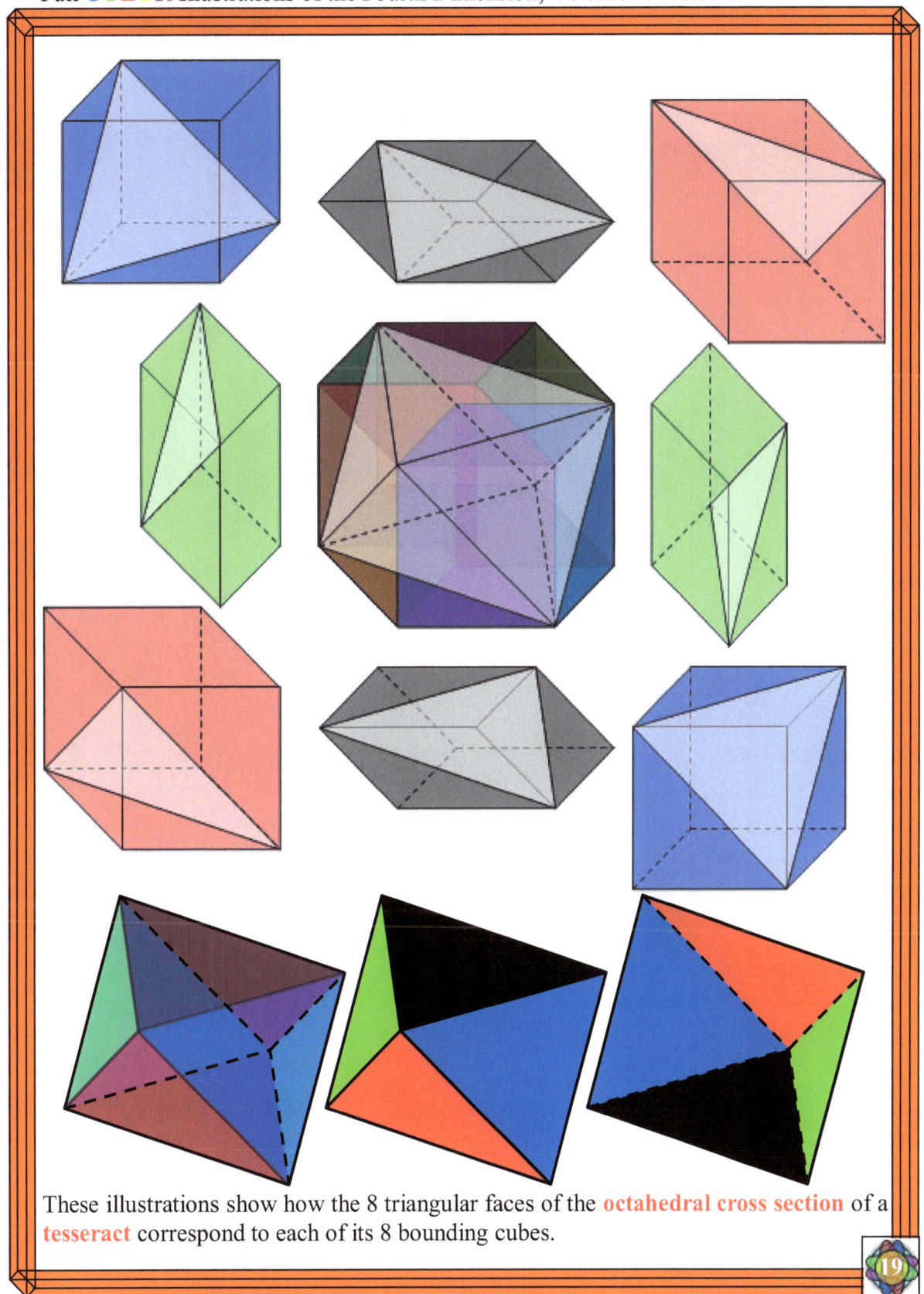

These illustrations show how the 8 triangular faces of the octahedral cross section of a tesseract correspond to each of its 8 bounding cubes.

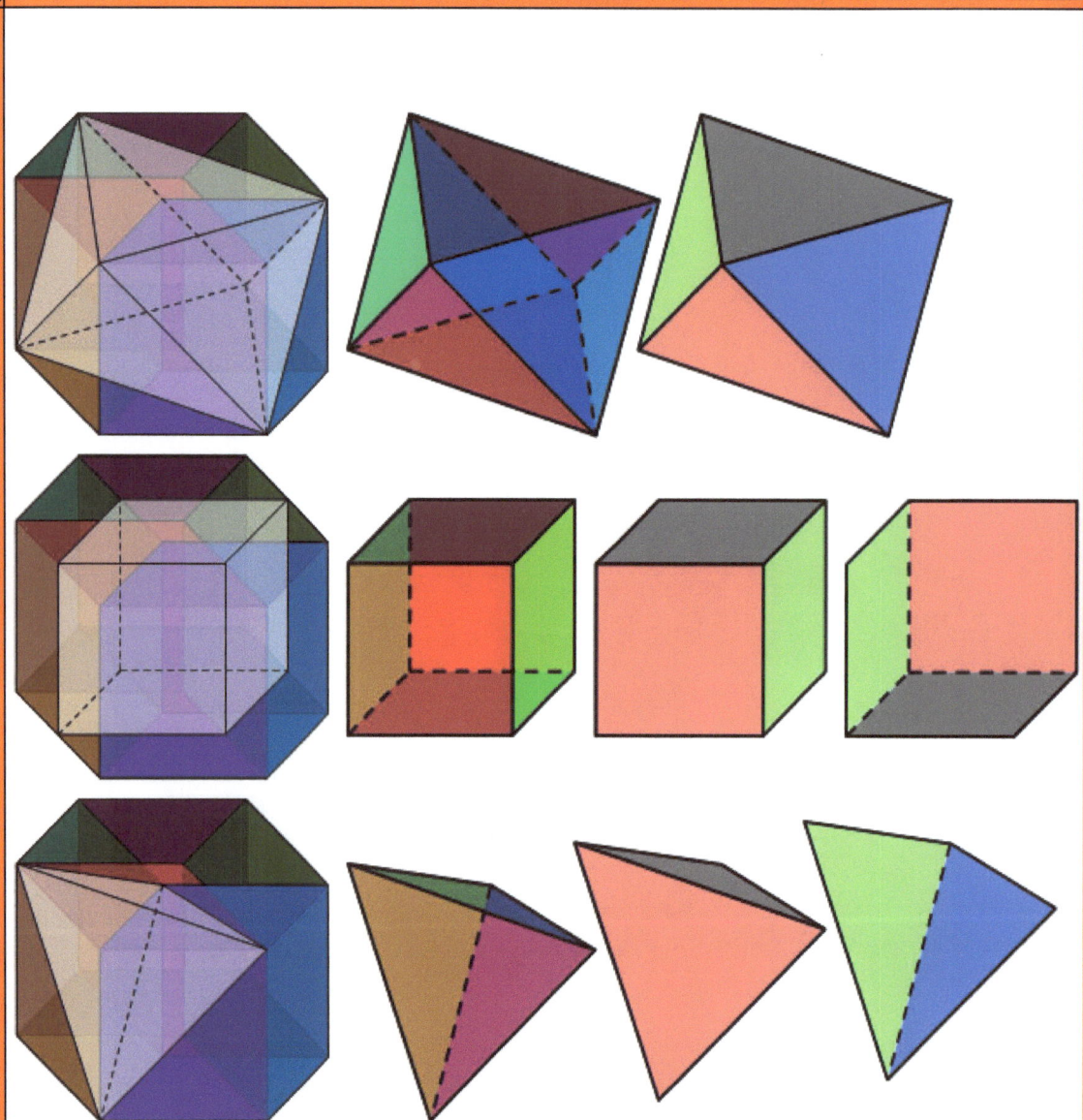

These **3D cross sections** represent the (3D) volumes of intersection that result from various hyperplanes (infinite cubes) with **tesseracts**. For the top diagram, the hyperplane is tilted such that it slices through the longest body diagonal of the tesseract; in the middle diagram, the hyperplane is parallel to the two blue bounding cubes (like slicing a block of cheese so as to bisect it into two rectangular blocks); and the hyperplane for the last illustration is parallel to a long body diagonal (a different diagonal compared to the first diagram), shifted closer to the top left corner. The resulting cross sections are the octahedron, cube, and tetrahedron, respectively. The side figures show which faces of the cross section lie in which of the 8 bounding cubes.

Full COLOR Illustrations of the Fourth Dimension, Volume 1: Tesseracts and Glomes

These 4D **hypercuboids** only differ from tesseracts in that the 4 types of edges do not all have the same length (but all edges of any given type do have equal lengths), just like the distinction between a rectangle and square. As with the tesseract, all edges, square (or rectangular) faces, and bounding cubes (or cuboids) intersect at right angles when they meet.

Part II: Glomes

(Hyperspheres in the Fourth Dimension)

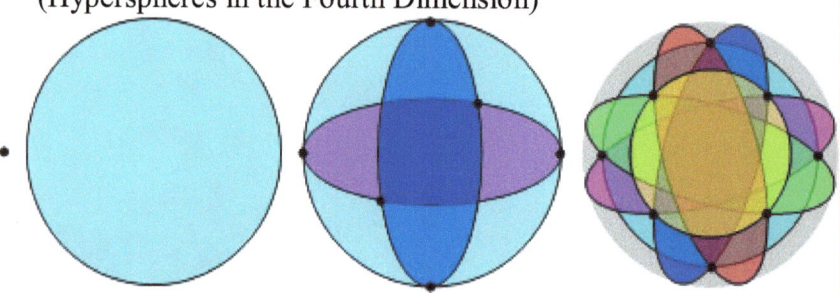

A **glome** is a 4D hypersphere. The hypersphere is a straightforward generalization of the sphere to N-dimensional space as follows: The **unit hypersphere** in N-dimensional space is the set of points one unit from the origin. The unit hypersphere is effectively $(N-1)$-dimensional in the sense that a particle constrained to lie on the hypersphere – i.e. constrained to be one unit from the origin – has only $N-1$ degrees of freedom, since it can not move radially inward or outward (toward or away from the origin) – just like a car on the surface of the earth can only move in two independent directions – north/south and east/west – even though the earth is three-dimensional. The equation for the unit hypersphere in N-dimensional space is $x^2 + y^2 + z^2 + w^2 + \cdots = 1$. Notice that the definition of the hypersphere does not include interior points (i.e. those points which are less than one unit from the origin). The geometric object defined to include the interior points is called the hyperball. A general hypersphere can be constructed by rescaling (moving each point the same distance from the origin) and translating (moving the hypersphere along a straight line) the hypersphere. Some lower-dimensional unit hyperspheres are:

- In 0D space, the hypersphere **does not exist**. (No point can be one unit from the origin).
- In 1D space, the unit hypersphere is the set of **two** 0D **points** one unit from the origin. Its equation is $x^2 = 1$ or $x = \pm 1$.
- In 2D space, the hypersphere is the 1D circumference of a **circle**. The equation for the unit hypersphere in 2D space is $x^2 + y^2 = 1$.
- In 3D space, the hypersphere is a 2D surface called a **sphere**. The equation for the unit hypersphere in 3D space is $x^2 + y^2 + z^2 = 1$.
- In 4D space, the hypersphere is a 3D hypersurface called a **glome**. The equation for the unit hypersphere in 4D space is $x^2 + y^2 + z^2 + w^2 = 1$.

Full COLOR Illustrations of the Fourth Dimension, Volume 1: Tesseracts and Glomes

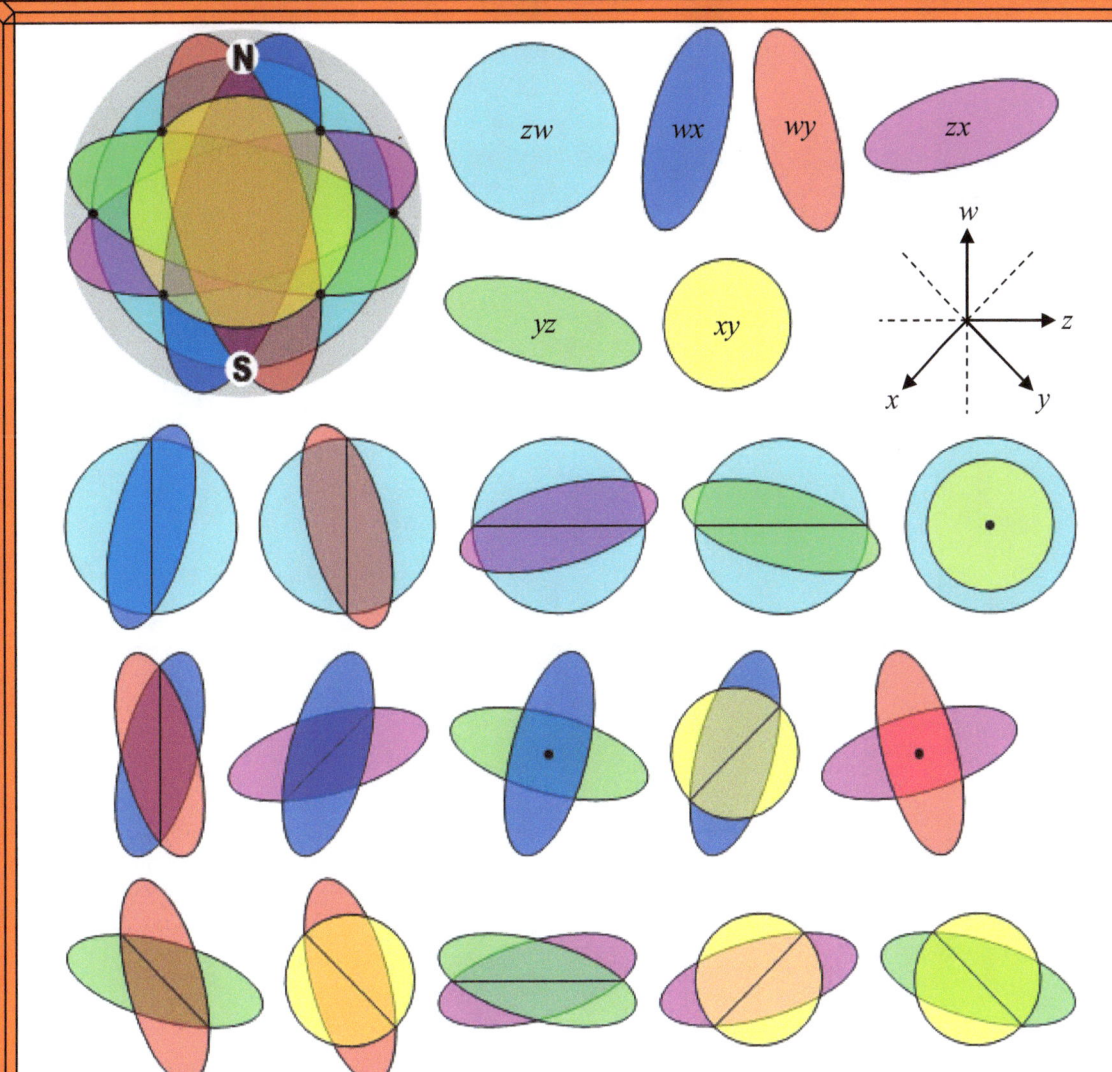

The **glome** (a hypersphere in 4D space) is a generalization of the sphere to the fourth dimension. It consists of the **3D hypersurface** that encloses the 4D volume inside. (A hyperball is a solid version of the hypersphere, which includes the interior volume.) **Six** mutually orthogonal **great circles** (a great circle is a circle with a radius equal to the radius of the glome) lying in the six mutually orthogonal Cartesian planes in four dimensions are illustrated for the glome above. The circumference of each circle lies in the hypersurface of the glome. The region of **intersection** between any two of these shaded great circles (making an exception this once to include their interiors) is a **line** if the planes of the circles have a mutual coordinate; otherwise, the region of intersection is a **point**. The eight points marked on the glome indicate where the great circles intersect (three intersect at each point). The gray shaded circle is not part of the set of mutually orthogonal great circles shown here, but represents the cross section of the glome; the glome can be rotated such that any one of the mutually orthogonal great circles shown could appear coincident with the gray circle (see later diagrams).

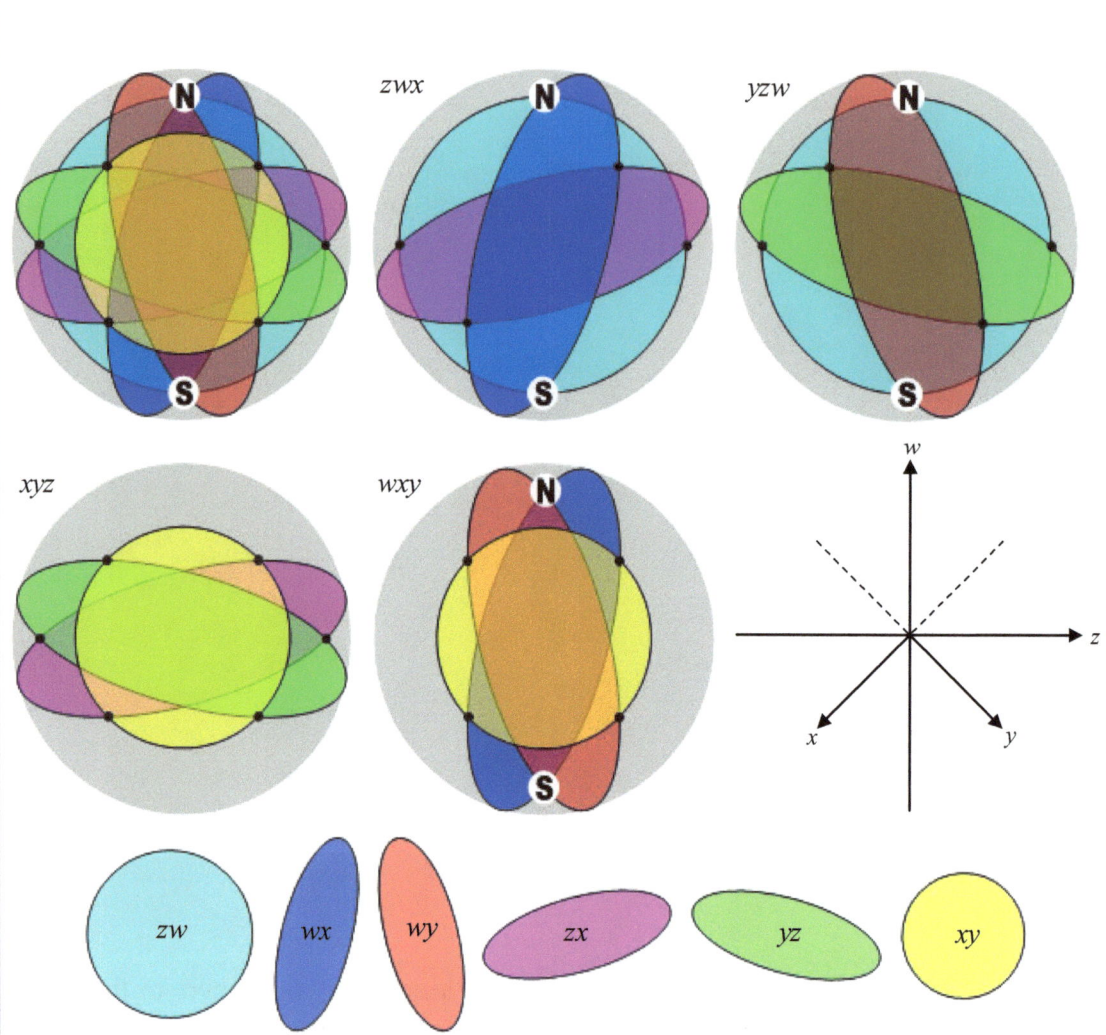

This glome is broken down into **four** mutually orthogonal **great spheres** (a great sphere being a sphere that has the same radius as the glome) – one corresponding to each of the four defining hyperplanes in four dimensions. The spheres – where a sphere includes only the two-dimensional surface (unlike ball, which is three-dimensional because it includes the interior volume) – lie on the hypersurface of the glome. The region of **intersection** between any two of these great spheres is one of the six mutually orthogonal **great circles** (the great circle lying in the plane of the two coordinates that are common to both spheres).

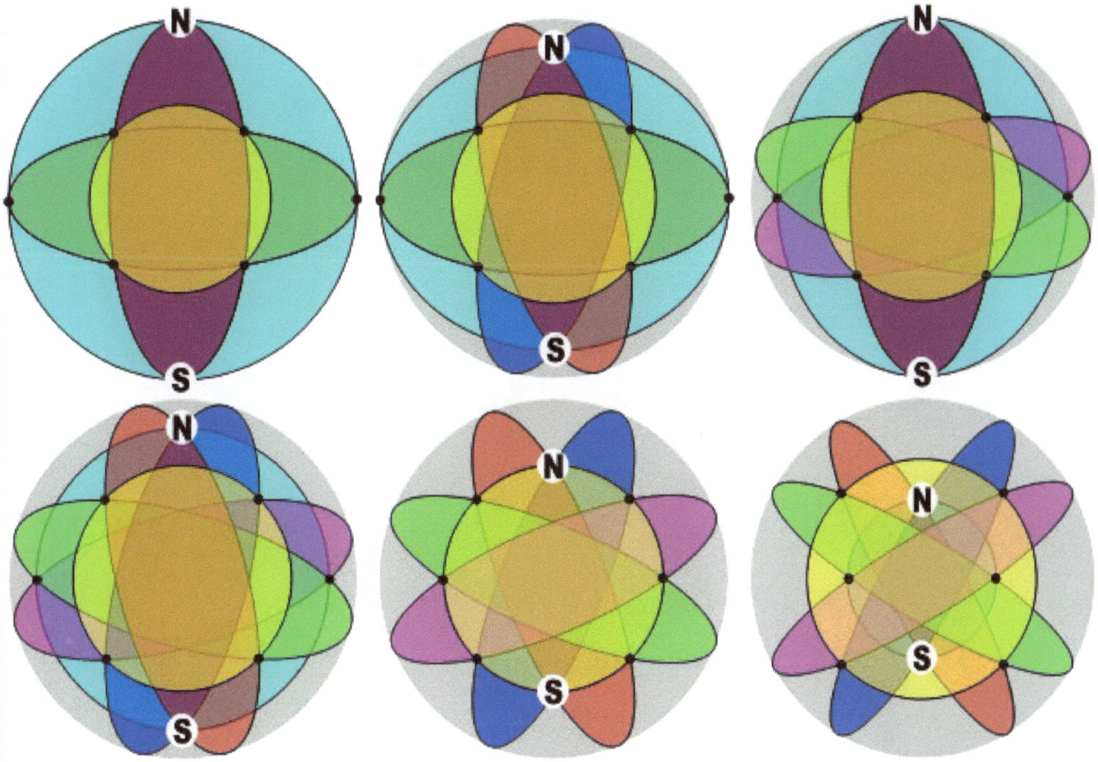

Full COLOR Illustrations of the Fourth Dimension, Volume 1: Tesseracts and Glomes

A uniform glome would look the same in the fourth dimension regardless of which way it was viewed, just as a uniform sphere looks the same from any angle in three dimensions. (A sphere, though, looks different from different perspectives in the fourth dimension, just like a circle looks different from various three-dimensional perspectives.) If the glome has one or more distinguishing features, however, then these features would look different from various **perspectives**. For example, if stripes were painted along a set of mutually orthogonal great circles, these stripes would change with viewing angle. Two pairs of planes have coincident projections on the two-dimensional plane of the page in the top left view. The next two views show the effect of rotating vertically or horizontally – e.g. in the top middle view, the north pole comes down on the front (and hyperfront) while the south pole rises in the rear (and hyperrear). The flowery diagrams at the bottom combine both senses of rotation. The projections of the *zw* and *xy* great circles are coincident from the symmetric perspective shown in the bottom middle illustration.

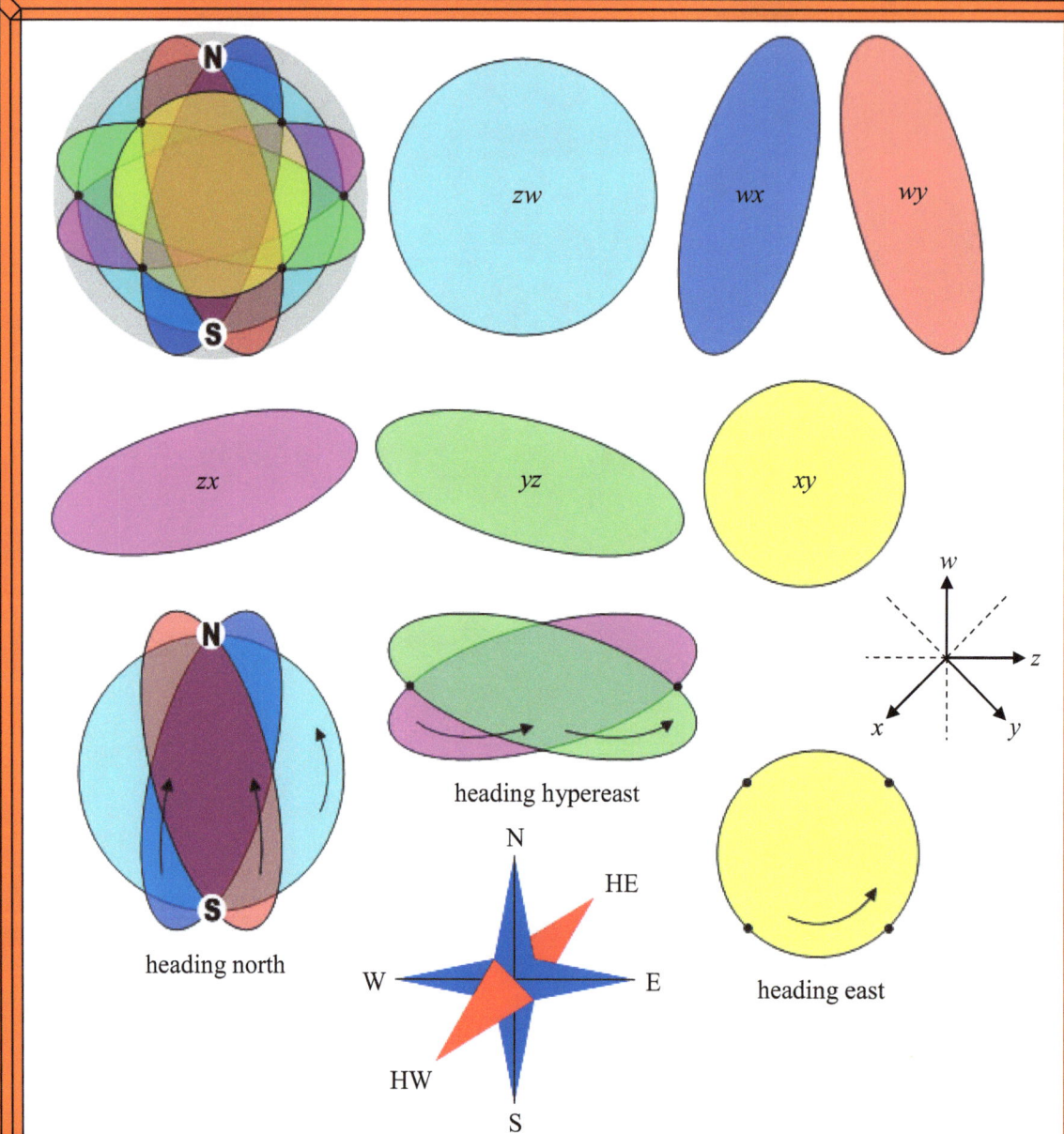

There are **three independent degrees of freedom** on the hypersurface of a glome: the three hyperspherical compass directions – **north/south**, **east/west**, and **hypereast/hyperwest**. North corresponds to traveling along a great circle of longitude that intersects the north and south poles, heading hypereast or hyperwest corresponds to a circle (generally not 'great') in the xy plane (for the coordinate system used above), and heading east or west corresponds to traveling in a circle (generally not 'great') perpendicular to the xy plane. When moving east, west, hypereast, or hyperwest, an object does not get any closer to either the north or south poles. Moving along any of the hyperspherical compass directions, an object does not get any closer to nor further from the center of the glome.

Full COLOR Illustrations of the Fourth Dimension, Volume 1: Tesseracts and Glomes

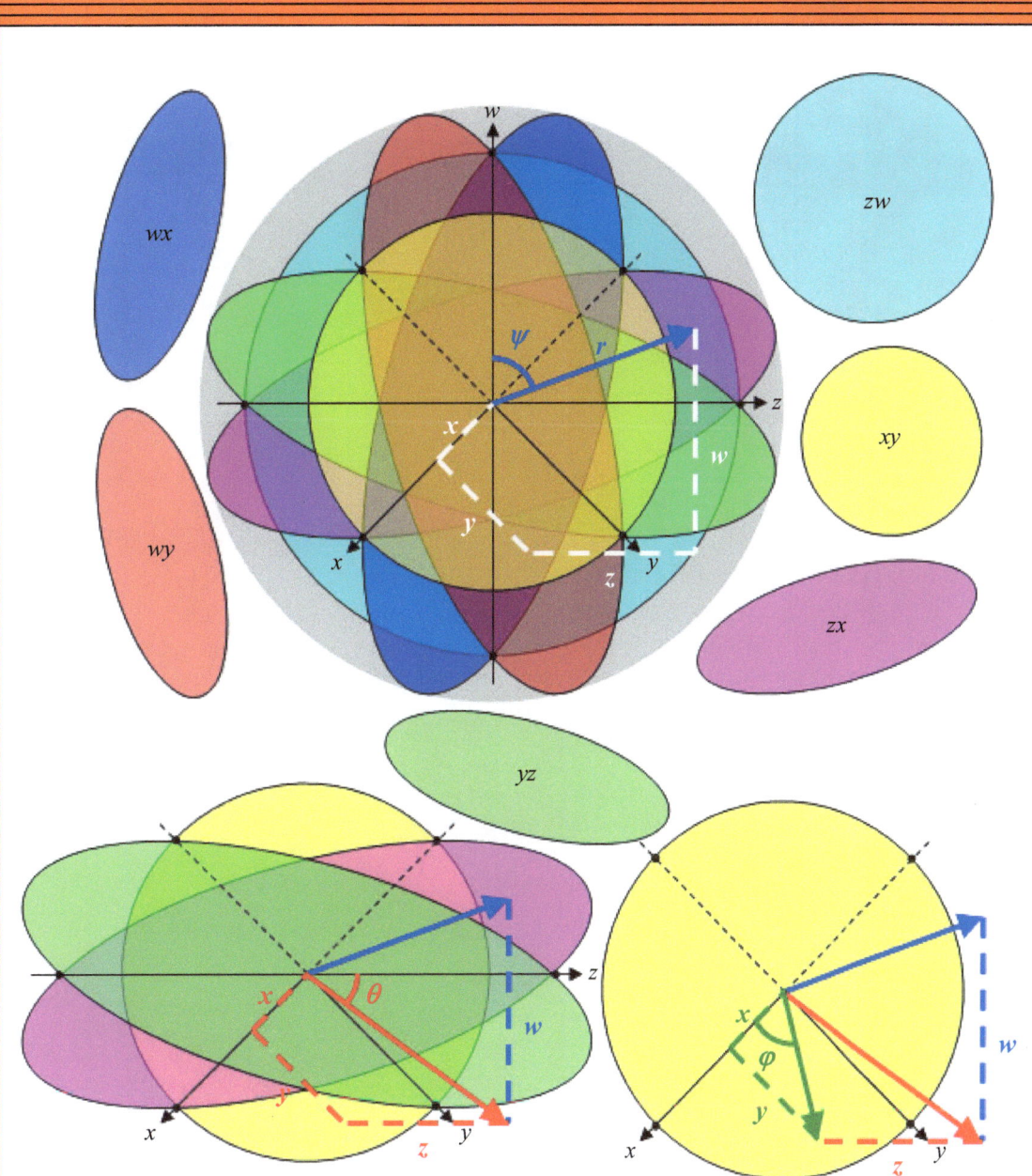

The **hyperspherical coordinates** r, ψ, θ, and φ form an equivalent alternative to the Cartesian coordinates x, y, z, and w. The **radial coordinate** r is the distance from the origin to a given point, the **polar angle** ψ is the angle r makes with the w-axis, the **hyperazimuthal angle** θ is the angle that r's projection onto the xyz hyperplane makes with the z-axis, and the **azimuthal angle** φ is the angle that r's second projection (i.e. after first projecting onto the xyz hyperplane) onto the xy plane makes with the x-axis.

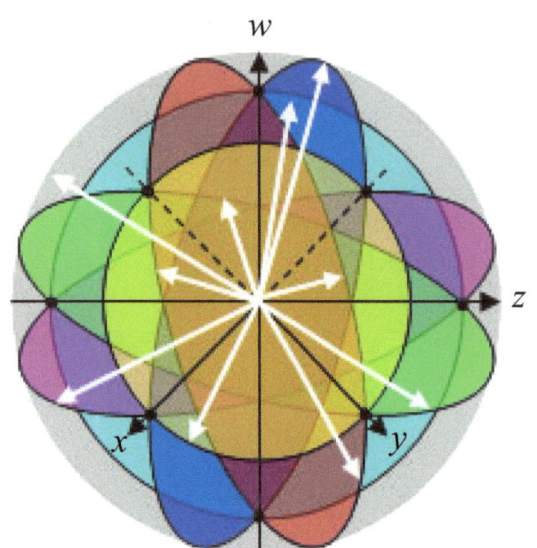

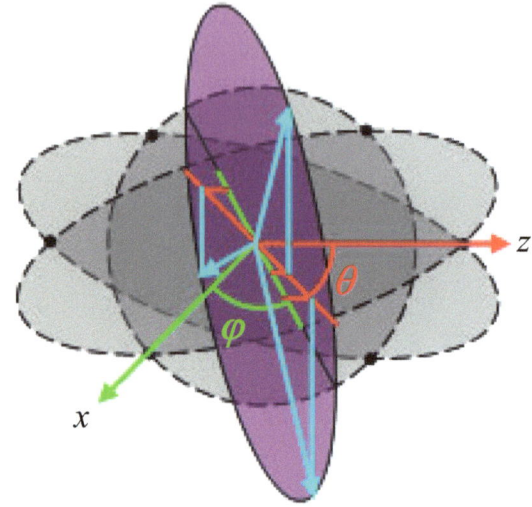

The **glome** is the set of points in 4D space that are equidistant from a common point in the center. Each white arrow on the diagram on the above has the same length and extends to a point on the hypersurface of the glome. Thus, a glome is defined by the equation $r = $ constant.

A **circle of longitude** is a great circle on the hypersurface of the glome that passes through the north and south poles. The red line represents its projection onto the xyz hyperplane (where $w = 0$), and the green line represents its projection onto the xy plane (where $z = 0$). Every point on the circle has the same two values for φ and θ.

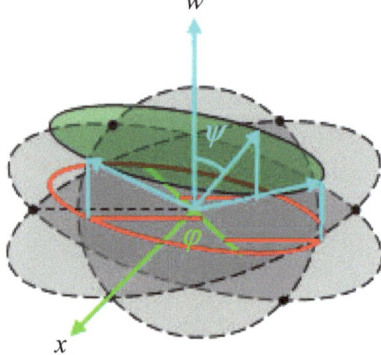

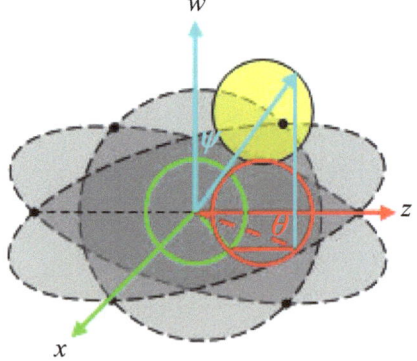

A **circle of hyperlatitude** lies in a hyperplane parallel to the **equatorial hyperplane**, but does not lie in a plane parallel to the xy plane. The red circle shows its projection onto the equatorial hyperplane (where $w = 0$) and the green line shows is subsequent projection onto the xy plane (where $z = 0$). Both φ and ψ are constants for a circle of hyperlatitude.

All **circles of latitude** lie in planes that are parallel to the xy plane. The image of the circle of latitude in the diagram above is projected downward along the w-axis until $w = 0$ (red circle) and then left along the z-axis until $z = 0$ (green circle). Both θ and ψ are constants for a circle of latitude.

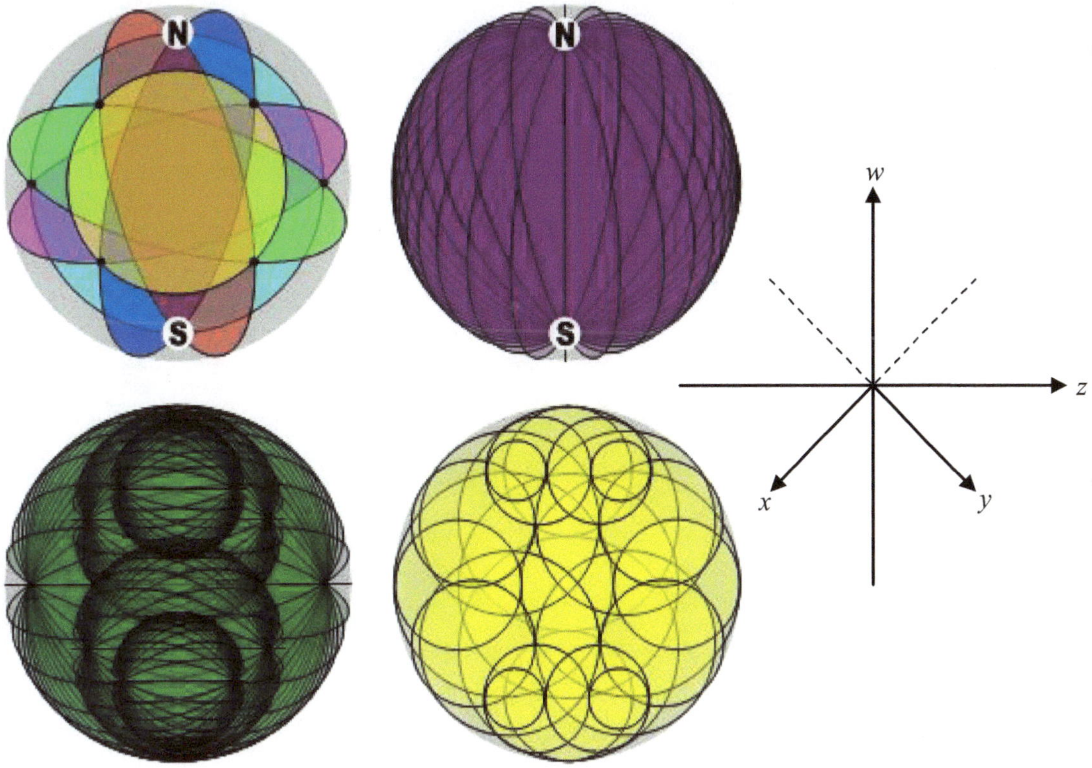

Full COLOR Illustrations of the Fourth Dimension, Volume 1: Tesseracts and Glomes

Circles of longitude (top right), **hyperlatitude** (bottom left), and **latitude** (bottom right) are illustrated here. All of the circles of longitude pass through both the north and south poles. Neither hyperlatitudes nor latitudes pass through these poles. For any of the five sets of hyperlatitudes shown, no circle of hyperlatitude is any closer to the north or south pole than any other circle of hyperlatitude from the same set. The large set of hyperlatitudes all lie on the equatorial sphere, where no point on the sphere is any more north or south than any other. All points on any circle of latitude lie in a plane parallel to the xy plane. Moving along a longitude corresponds to north/south, moving along a hyperlatitude corresponds to hypereast/hyperwest, and moving along a latitude corresponds to east/west. No two hyperlatitudes or two latitudes intersect, and longitudes only intersect at the poles. (There are actually more circles of longitude on the glome than there are circles of hyperlatitude on the equatorial sphere: However, because of the symmetric projections along x and y from the perspective shown, many of the circles of longitude are coincident. Circles of longitude include great circles in both the zwx and yzw hyperplanes, but for every longitude in zwx there is a coincident longitudinal projection in yzw. All circles of hyperlatitude, on the other hand, lie in a xyz hyperplane, and so do not suffer from such projective degeneracy.)

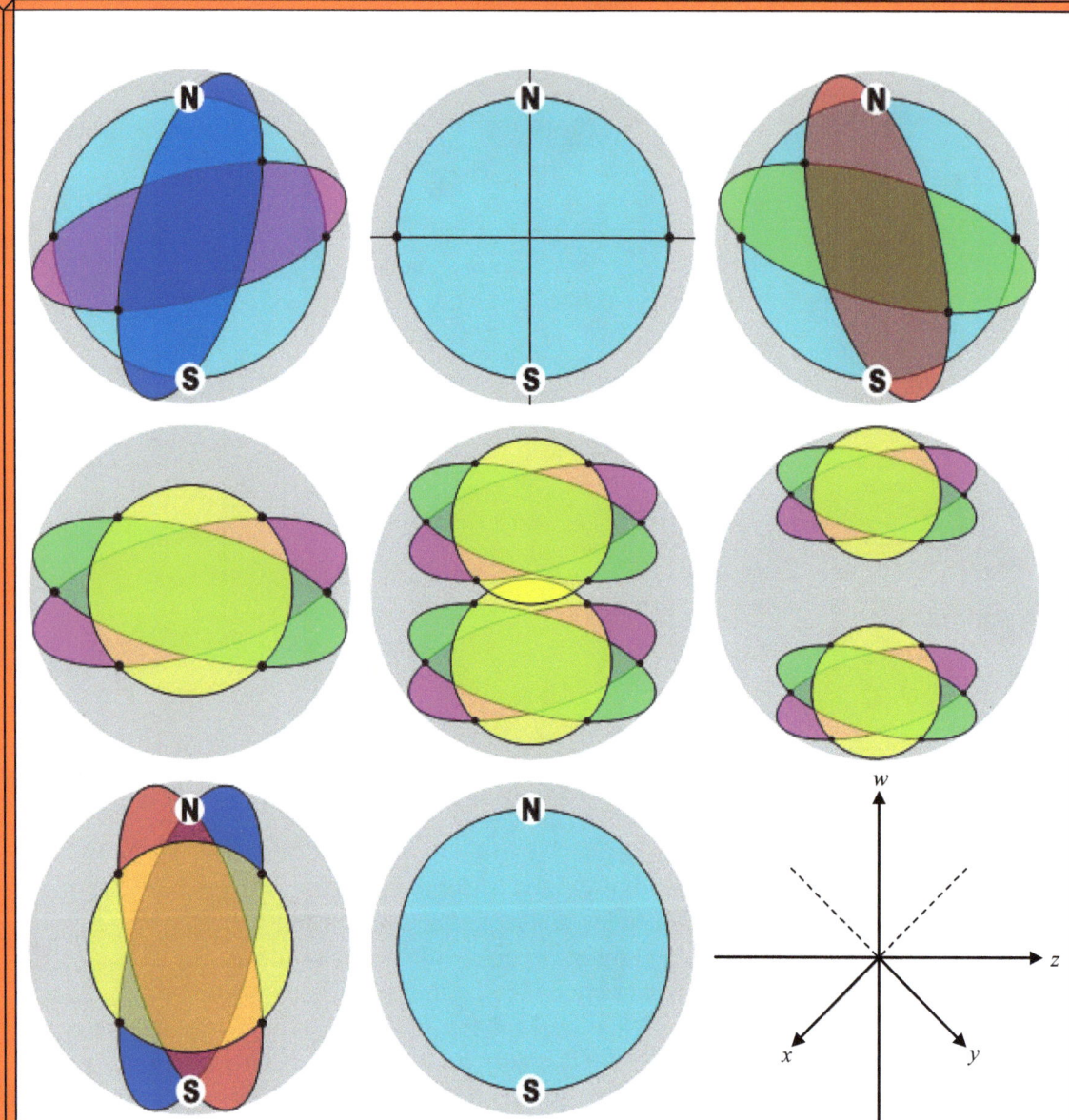

Spheres of longitude (top) correspond to $\varphi =$ constant and are great spheres (radius equal to that of the glome). These include the great zwx (left) and yzw (right) spheres, as well as any great sphere uzw where u is a linear combination of x and y. The central sphere of longitude shows a perspective for which two of its mutually orthogonal great circles are horizontal and vertical. **Spheres of latitude** (middle) correspond to $\psi =$ constant. No part of a sphere of latitude is any closer to the poles than any other part. The left diagram shows the equatorial sphere. **Surfaces of hyperlongitude** (bottom) correspond to $\theta =$ constant. This is the great wxy sphere for $\theta = 90°$, but rotates and flattens into the great zw circle as θ approaches $0°$ or $180°$.

Full COLOR Illustrations of the Fourth Dimension, Volume 1: Tesseracts and Glomes

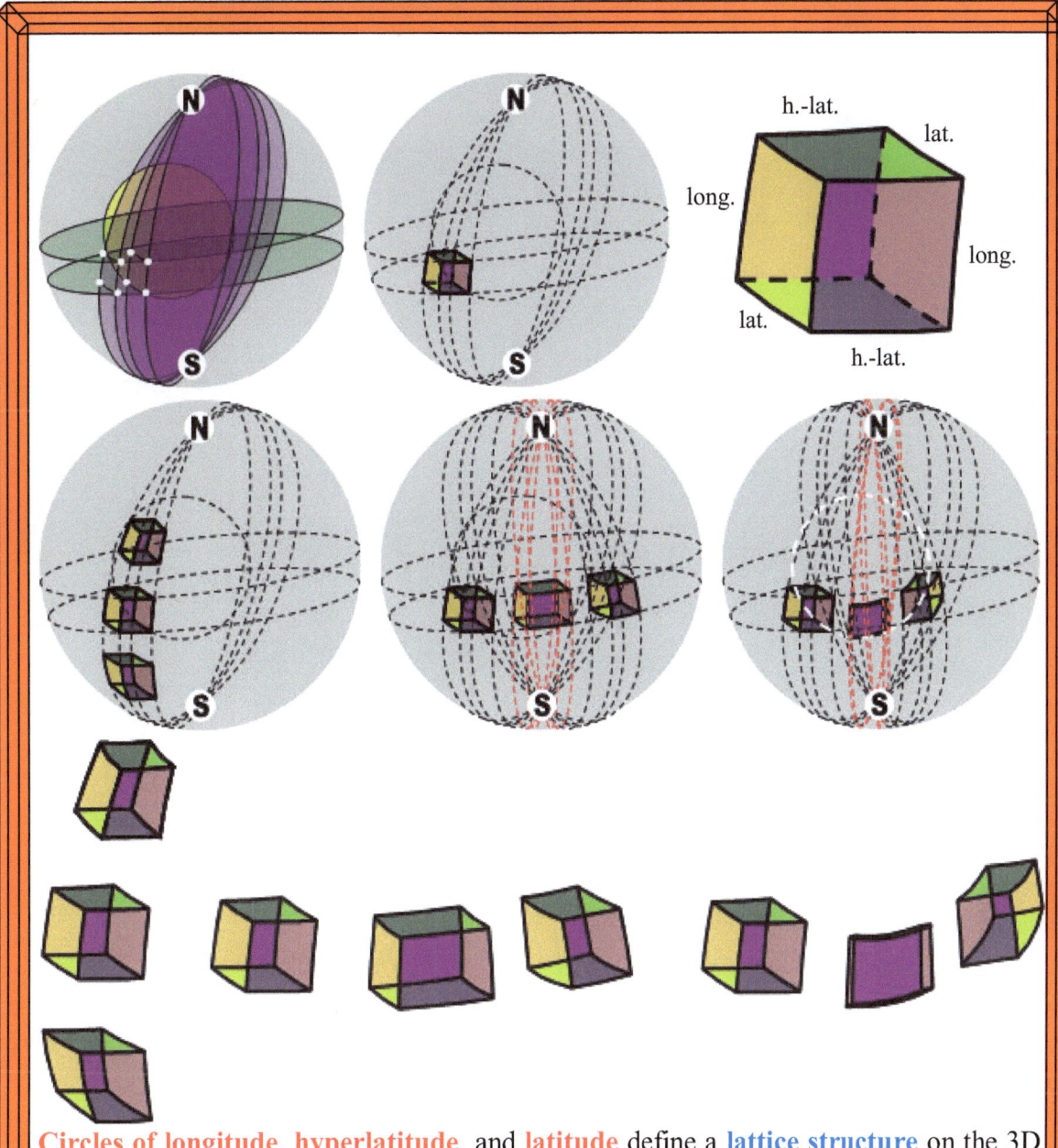

Circles of longitude, hyperlatitude, and latitude define a **lattice structure** on the 3D hypersurface of the glome. An object on the surface of a sphere can travel north, east, south, and west, returning to its starting position along a warped square – i.e. its "sides" are circular arcs. On the hypersurface of a glome, an object can also make similar squarish tiles traveling north, hypereast, south, and hyperwest, or hypereast, east, hyperwest, and west. These three mutually orthogonal tiles define a cube-like lattice structure with circular edges and spherical sides. No part of any cube (interior included) lies closer to the center of the glome than any other part. Sets of 3D tiles are shown for a longitude (middle left), hyperlatitude (center), and latitude (middle right). A complete set of 3D tiles would form a simple dimple pattern for a 4D golf ball.

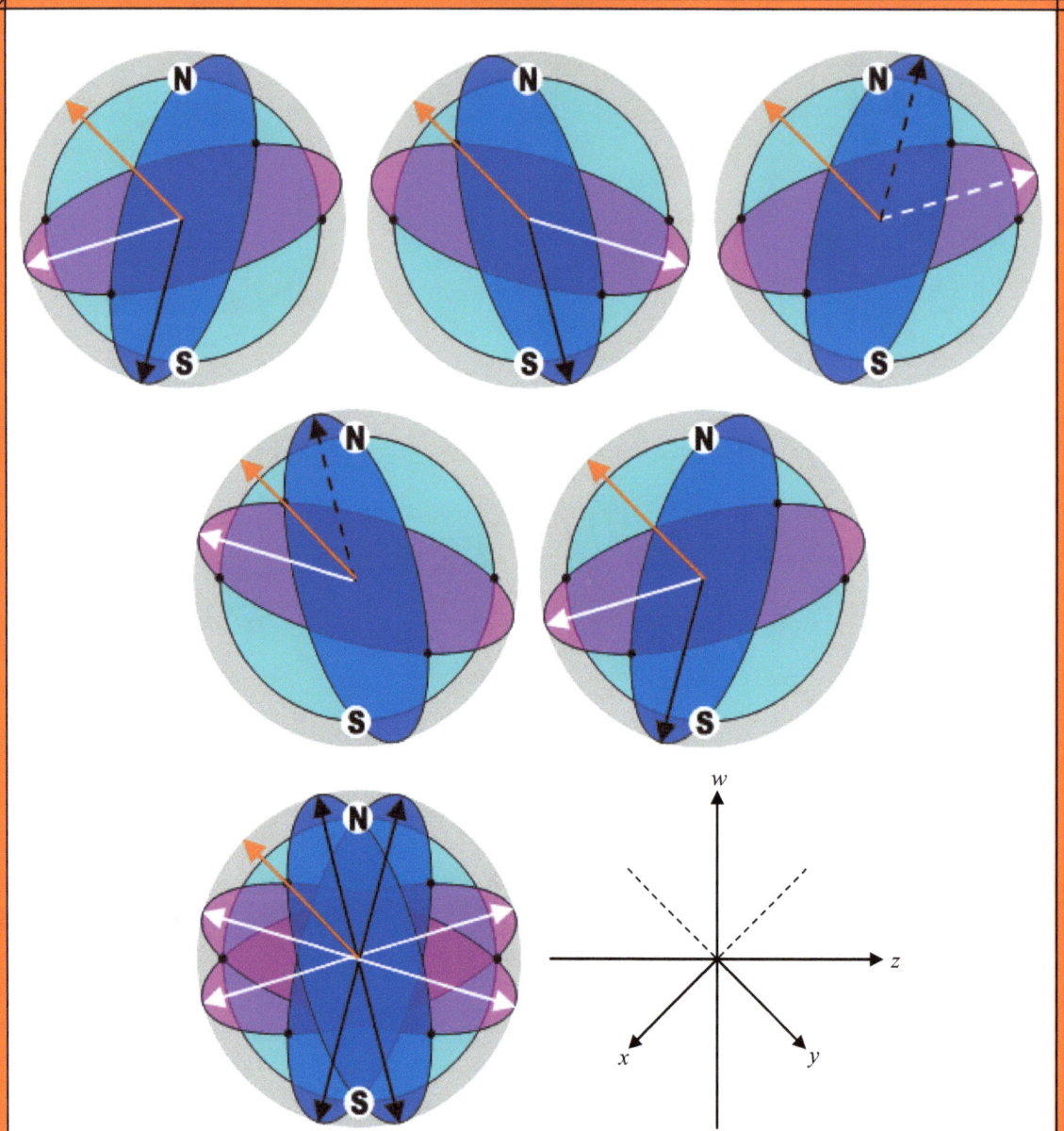

This **sphere**, initially lying in the *zwx* hyperplane (a 3D subspace), **rotates** about the *zw* plane. Each rotation shown corresponds to a 90° rotation in 4D space. Each 90° rotation transforms the sphere between the *zwx* and *yzw* hyperplanes; these two hyperplanes are mutually orthogonal and intersect at the *xy* plane (there is no other overlap). The great circle lying in the *zw* plane is unaffected throughout this rotation, while the *zx* and *yz* great circles and, similarly, the *wx* and *wy* great circles, rotate into one another through each 90° rotation. The great circles *zx*, *yz*, *wx*, and *wy* **reflect** upon each **180° rotation**. The bottom diagram is a composite image of the rotated spheres put together.

Full **C**O**L**O**R** Illustrations of the Fourth Dimension, Volume 1: Tesseracts and Glomes

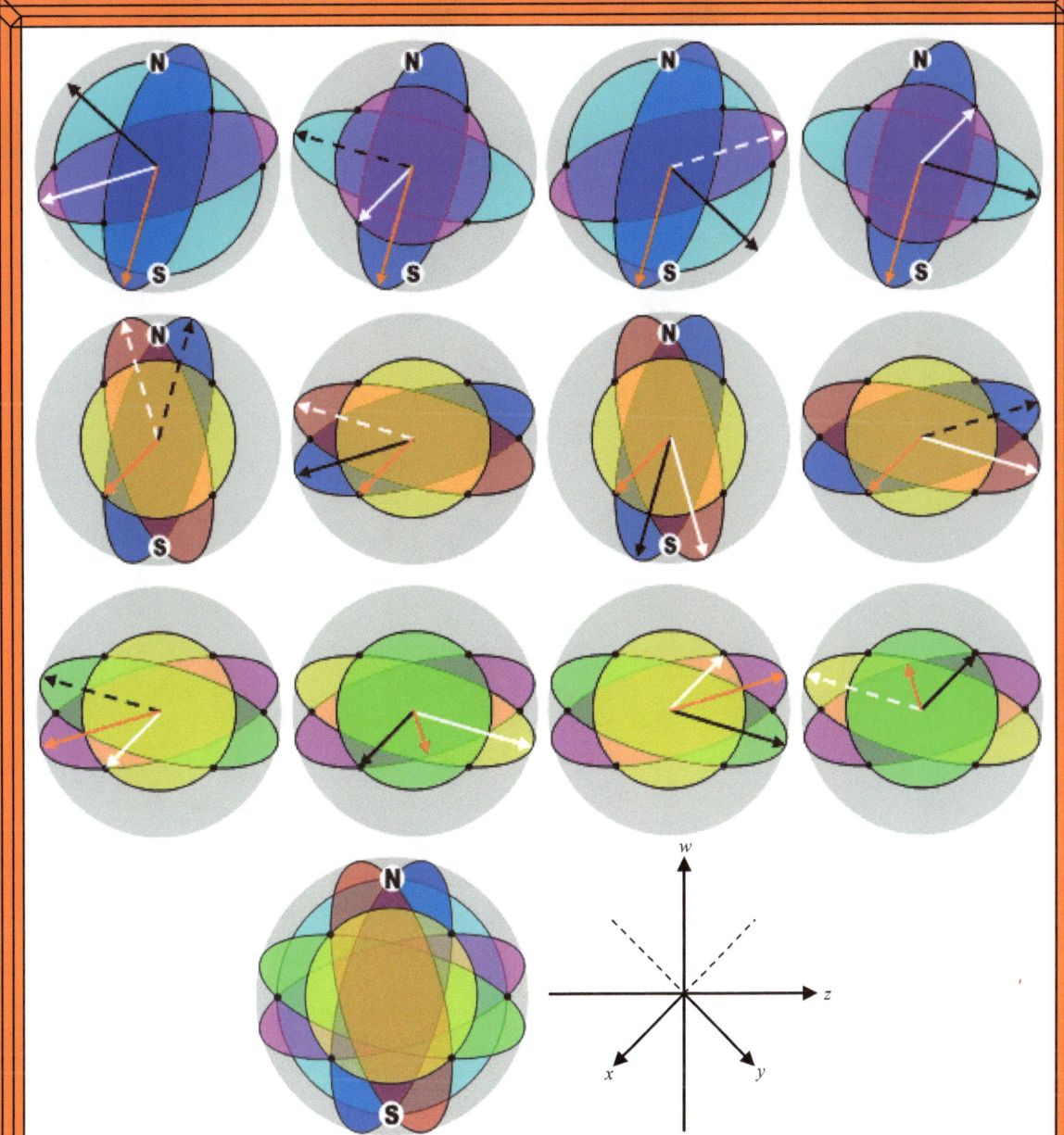

The top diagrams show a **sphere rotating** about the *wx* plane, the middle diagrams show a rotation about the *xy* plane, and the bottom diagrams show a rotation about the *yw* plane. The first two rotations are fully four-dimensional, rotating about a plane that includes one of the rotating sphere's great circles. In these cases, the great circle perpendicular to the plane of rotation remains stationary while the other great circles rotate through the 4D space. The last rotation is effectively 3D as the plane of rotation is orthogonal to the rotating sphere (the distinction equates to just one axis, not two, of the rotating sphere being in common with the plane about which the sphere rotates). The sphere does not reflect upon a 180° rotation in this latter case (i.e. it does not turn a right-handed object left-handed; note that the projected images can be illusory).

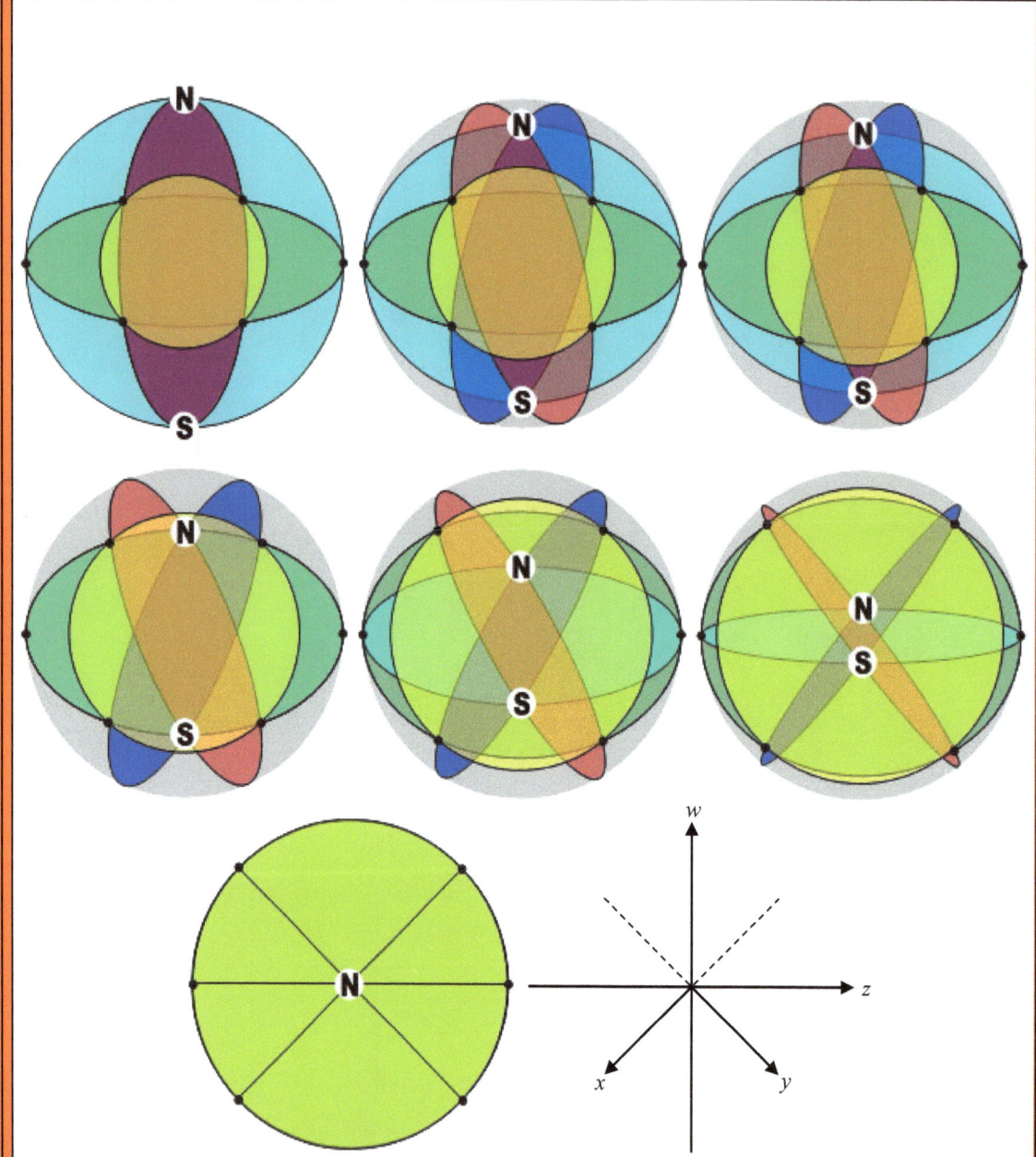

This **glome rotates** about a plane that includes the z-axis and an axis that bisects the x- and y-axes. From this symmetric perspective, the projections of the zx and yz great circles are coincident. In the final position shown, the projections of three of the great circles have collapsed down to line segments (diameters), which are all perpendicular, of course. Similarly, from this symmetric perspective, in the final projection three of the planes appear to be coincident, yet remain mutually orthogonal.

Full COLOR Illustrations of the Fourth Dimension, Volume 1: Tesseracts and Glomes

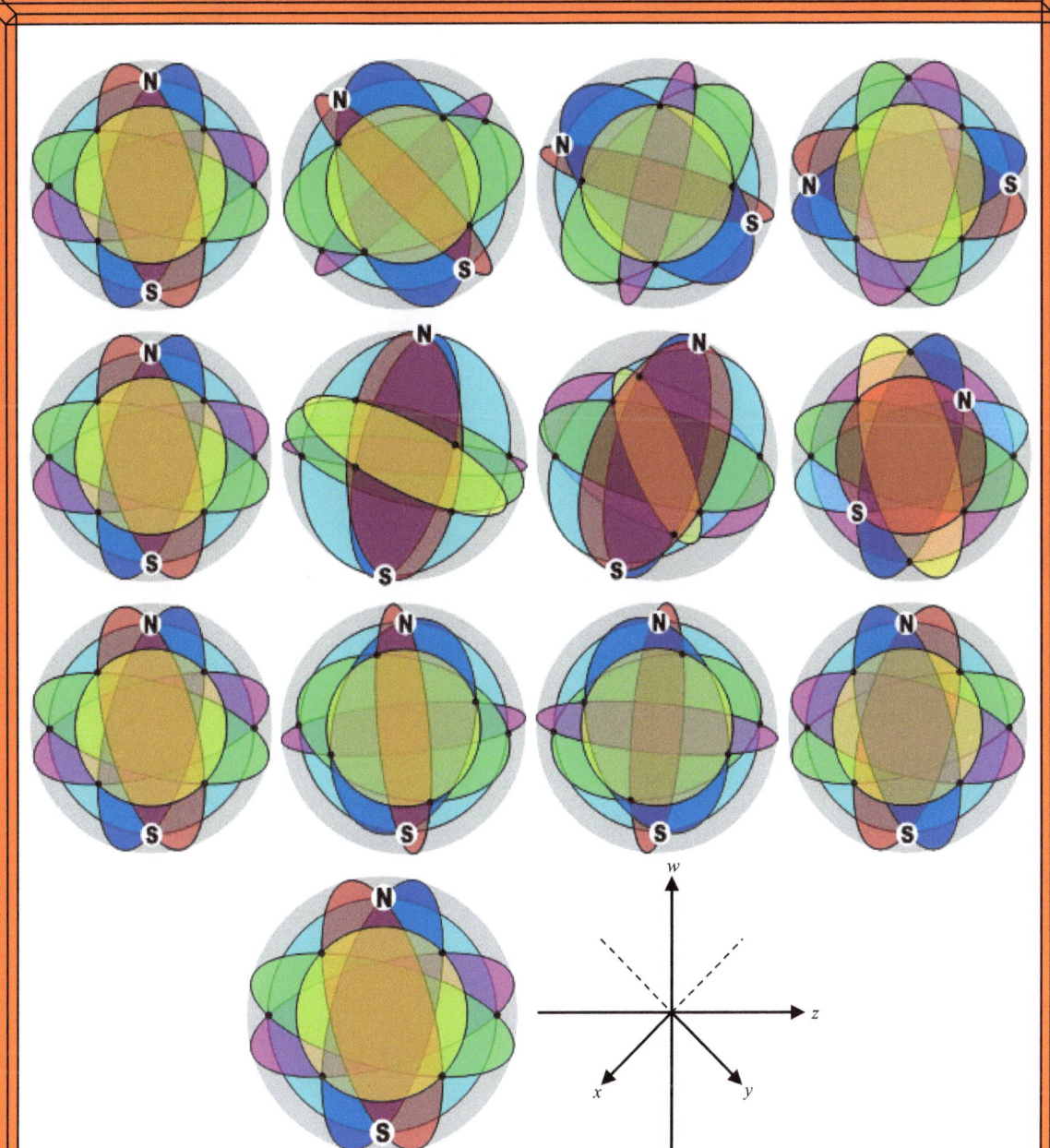

The top diagrams show a **glome rotating** about the xy plane. In this case, the zw great circle rotates counterclockwise in the top rotation, the xy great circle remains stationary, and the two other pairs of great circles rotate so as to swap position through 90°. The glome in the middle diagrams rotates about the yz plane. This requires the yz great circle to remain stationary, the wx great circle to simply turn on its axis, and the other two pairs of great circles to rotate through 4D space so as to swap position. The bottom diagrams show a rotation about the yw plane. Its great circles are analogously affected.

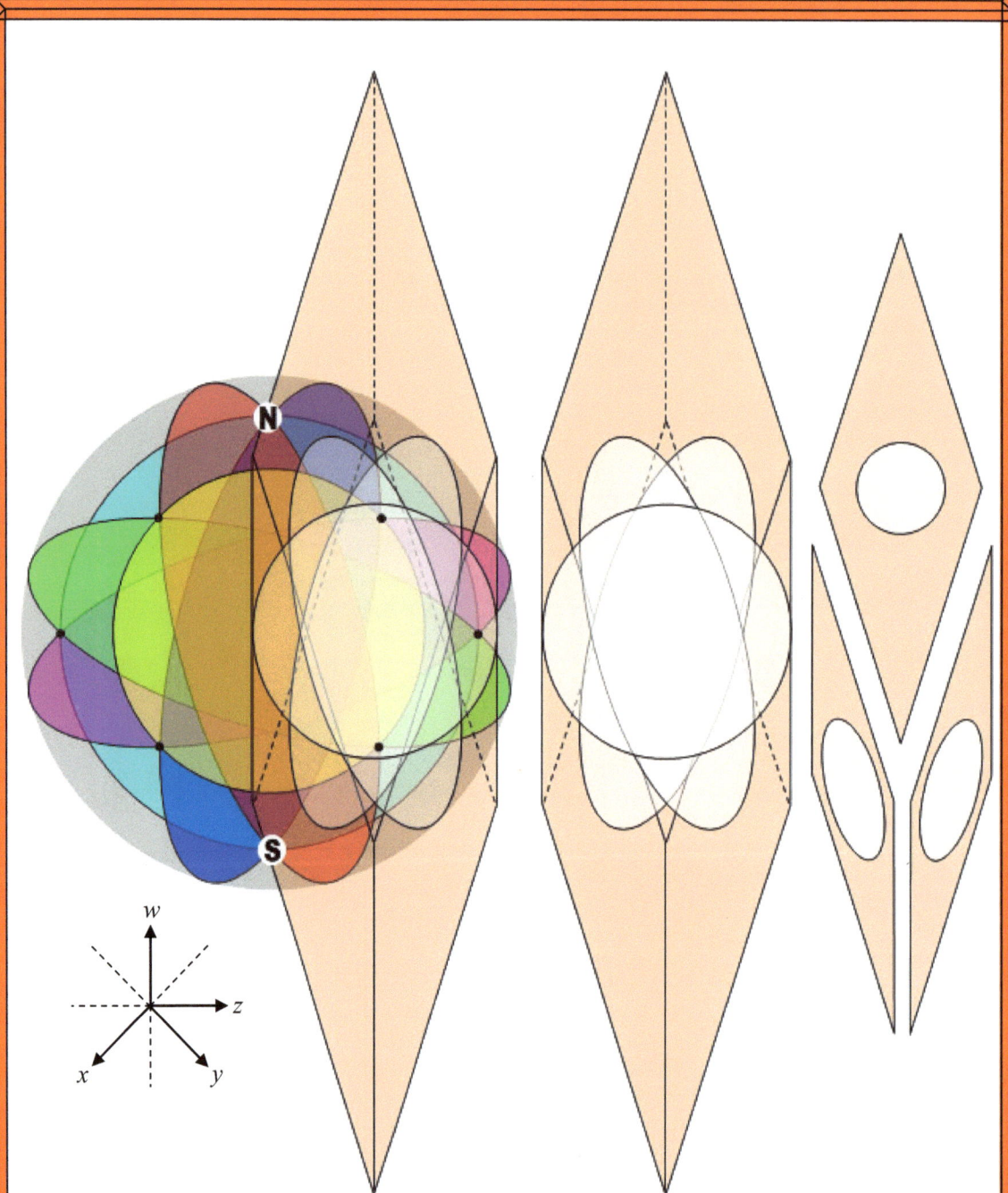

The **intersection** of this **glome** and the *wxy* **hyperplane** is a **sphere** (in general, not "great") in the *wxy* hyperplane (the set of points on the glome for which *z* has the same value as the hyperplane). This sphere has infinitesimal thickness in the *z* dimension; it is orthogonal to the *z*-axis. The diagram on the right shows the circular cross sections that result when the three planes orthogonal to *z* slice the glome; these planes form the six sides of the *wxy* hyperplane.

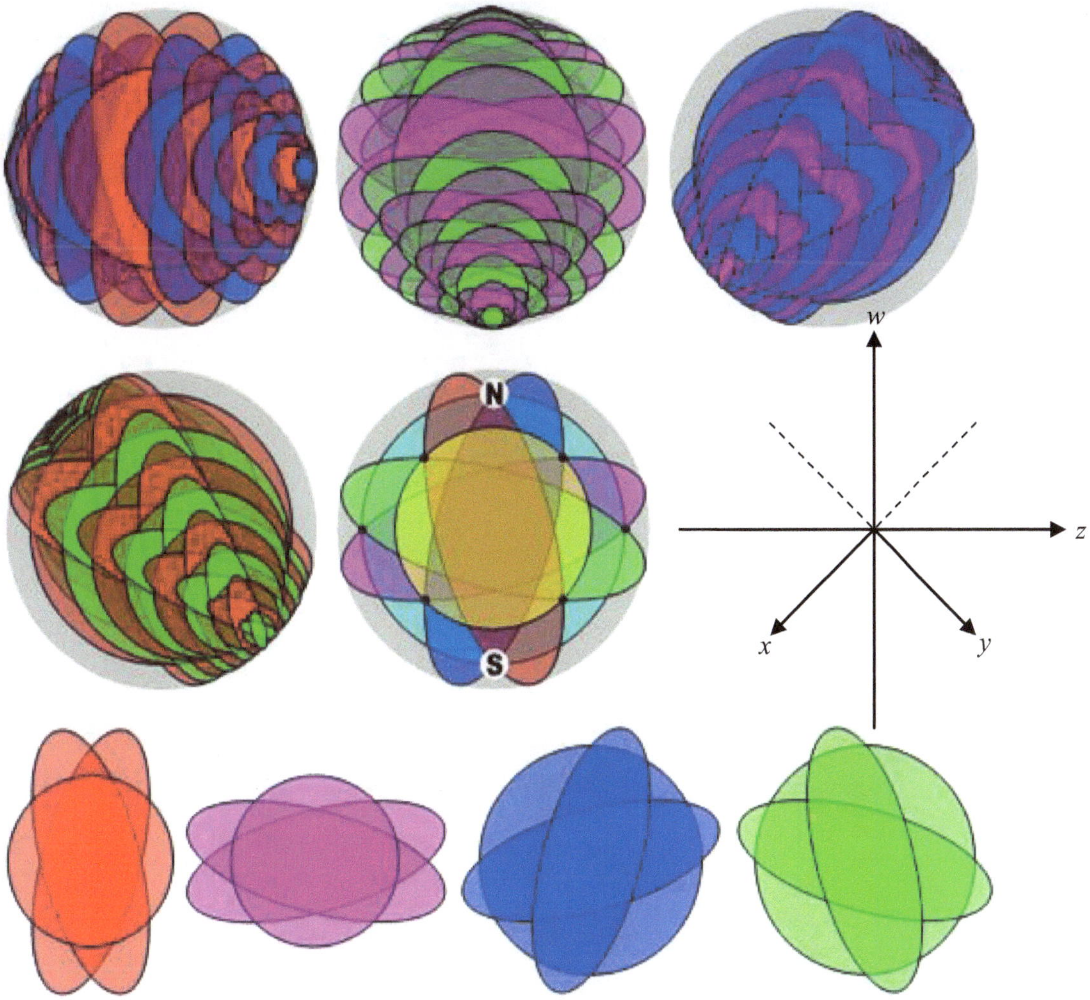

Full COLOR Illustrations of the Fourth Dimension, Volume 1: Tesseracts and Glomes

Each of these **glomes** is **sliced** into several **parallel sub-spheres**. The sub-spheres are infinitesimally thin compared to the glomes. No two sub-spheres overlap or touch. The centers of the sub-spheres lie on the diameter of the glome that is perpendicular to the sub-spheres. This is what it would be like to chop a hyperonion with a hyperknife in 4D space. The four types of sub-sphere slices that comprise the slicing diagrams are shown individually at the bottom. These correspond to the four fundamental hyperplanes of 4D space, and the four mutually orthogonal spheres of a glome.

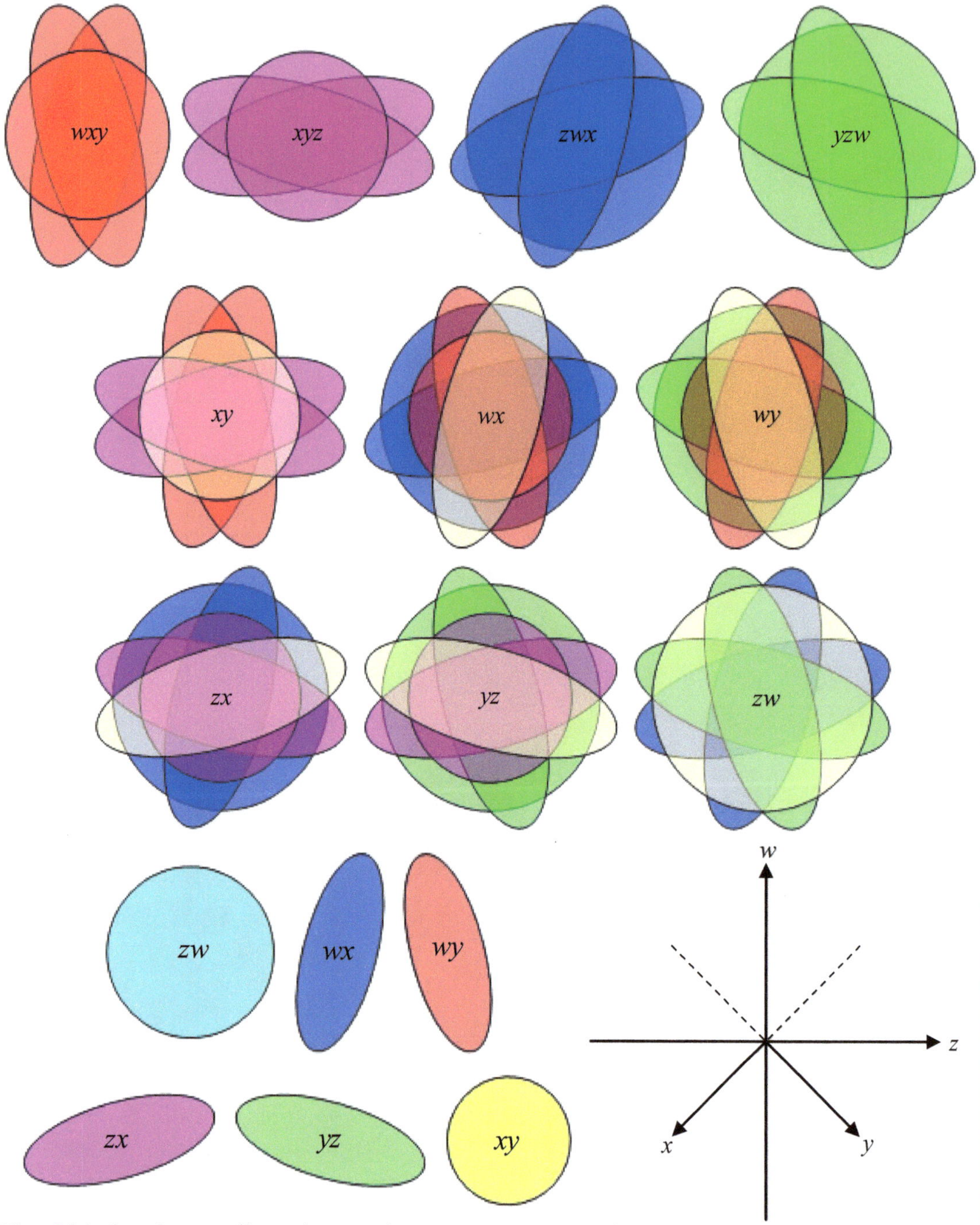

The **4** kinds of mutually orthogonal **great spheres** can intersect **6** distinct ways. In each case, the region of **intersection** is a **great circle** lying in the plane of the two shared coordinates (e.g. the *wxy* and *yzw* great spheres intersect at the *wy* great circle).

Full **COLOR** Illustrations of the Fourth Dimension, Volume 1: Tesseracts and Glomes

A **hyperellipsoid** is like a hypersphere that is distorted by scaling one or more of the Cartesian coordinates $(x, y, z, w, \ldots)$ with different factors. The equation for a hyperellipsoid in 4D space is $(x/a)^2 + (y/b)^2 + (z/c)^2 + (w/d)^2 = 1$. The equation for a **glome** is obtained by setting $a = b = c = d$. With three equal coefficients, a **prolate** or **oblate hyperglomoid** results if the unequal coefficient is greater or less than the other coefficients, respectively. Various types of **hyperspheroids** result from two equal coefficients.

The second volume in this series will apply these fundamental geometric structures to illustrate **4D objects in a 4D world**, analogous to common 3D objects with rectangular or spherical shape. This includes:
- Stacking tesseracts to build a hyperpyramid, as ancient Egyptians in 4D
- Building and rotating a 4D version of a Rubik's cube
- Generalizations of common objects like a chair, table, or cross
- What it would be like to live in a hyperhome
- The lattice structure of hypercrystals and supermarket packing
- Geography and astronomy in 4D space
- A sample alphanumeric system for arithmetic and writing
- Reflections from a hyperplane or hyperspherical mirror

*Full **COLOR** Illustrations of the Fourth Dimension, Volume 2* is scheduled for release in the **fall of 2009**. The author's other works on the subject of the fourth dimension include:
- *The Visual Guide to **Extra Dimensions**, Volume 1: Visualizing the Fourth Dimension, Higher-Dimensional Polytopes, and Curved Hypersurfaces*.
- *The Visual Guide to **Extra Dimensions**, Volume 2: The Physics of the Fourth and Higher Dimensions, Compactification, and Current and Upcoming Experiments to Detect Extra Dimensions*.

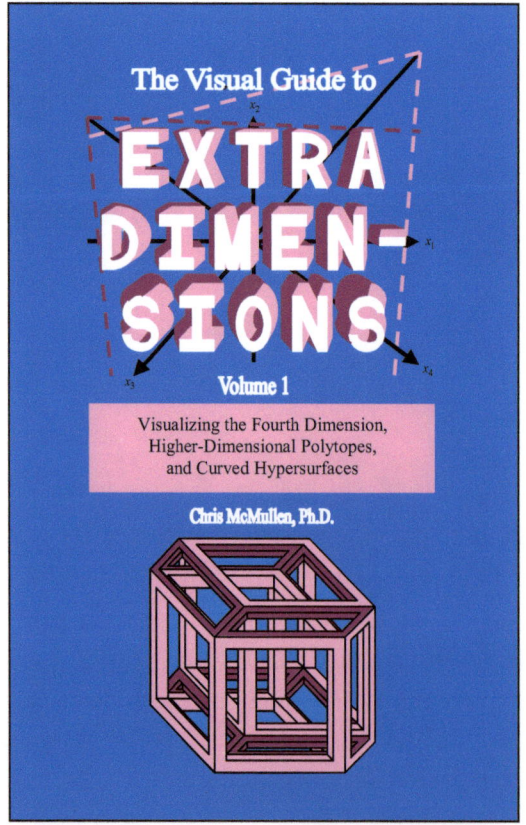

www.ingramcontent.com/pod-product-compliance
Lightning Source LLC
Chambersburg PA
CBHW051107180526
45172CB00002B/808